I0486860

From Electric Numbers to Hal Trees

Also by Harold L. Reed

Brains for Machines/Machines for Brains
Neurons for Robots

Copyright 2009 Harold L. Reed

ISBN 9781441422798

Math, Algebra, Computers, Microcontrollers, Neurons

From Electric Numbers to Hal Trees

By
Harold L. Reed

Table of Contents

Table of figures

Preface

There are several mathematical holes, aggravations.
and overlooked structures that I want to fix. Mathematics is a language whose grammar is mostly, to my ignorant knowledge, correct, but most of that was set up before the computer was invented.

One aggravation is nomenclature. Single symbol variables are hard to read. a = bc means a = b*c. Why not say that. Introduce the * as a multiplication operator and you can have long variable names. One must distinguish against functions and subscripts. F(x) for function and A[1] for index for example. Math texts should be readable by a machine. Comments in the text can follow ;. All this comes from the computer.

One thing I want to add is to put the computer into mathematics, not as a calculator, but as structure. Then we can replace functions with computers.

One overlook is mechanizing data streams. I provide an algebra of data streams. Data streams are written and read by computers.

Math has nothing like Hal trees. I introduce and play with them for you. Hal trees are networks of computers.

All of these are combined in this book. Each is its own little book.

book 1
What is First?

To create a language you need a language. To make a mathematical language one must use a less precise language, and then rewrite the language in its own language.

We need no history here except for flavor. It is hard to know where to start, since we have history before us, but I want to start at the beginning, not in historic time, but in essence.

We must start with a single mathematical object and build from that. Take a number symbol as an object and build on that. Take **ThisObject** as a symbol for an unspecified number. ThisObject becomes a variable (which I will define later.) Note that a variable holds the place for the number, representing it but it is not the number. When we say **ThisObject** is **2** then we have given it a value, but it is still not the number.

A language is a system of actions, objects and the rules of actions on objects. Given a language, one can make definitions of objects, actions and actions on objects.   When you run out of definitions you are left with objects that are primary and not definable.

Variables and Sets

A variable is a set with one object. You make a variable by giving it a name. ThisObject could be a variable name.

A set is a container of objects. A set is made by naming it. ThisSet is a set name.

Equality and Containment

To make an equation you must invent or take an equal sign which is equivalent to 'is' in spoken language. We use = so we can say ThisObject = ThatObject. This means that ThisObject can be replaced by ThatObject anywhere we wish.

ThisObject and ThatObject are variables, which are sets which are containers. Containers are boxes that contain variables or sets and maintain their positions or place in whatever system they are in.

Systems and Language

A system is a set of objects and actions with a list of possible actions on the objects of the system.

s = (objects, actions, program)

All languages are systems containing nouns and connecters with verbs for action and grammar for the program of actions with the objects and connecters.

Objects and Actions

Since a language is a system of
objects and actions, we need to see which
is which in a mathematical system.

I say **ThisObject = ThatObject +
AnotherObject**. ThisObject, ThatObject, and
AnotherObject are objects here and the =
sign and the + sign are the action. The
action is to add ThatObject and
AnotherObject and store the result, as a
copy, into ThisObject.

ThisObject follows ThatObject and
AnotherObject in time. In another sense,
AnotherObject + ThatObject cause
ThisObject. ThatObject and AnotherObject
define ThisObject.

Objects

Some objects of this language are:
Sensors are to sense some attribute of
outside
Nodes are functions.
Motors push on outside.
Numbers are the codes for Sensors, Nodes
and Motors.
...

A variable is a named object or is the
name of an object. An object can have
multiple names.

Properties of Objects

Every object has its own properties. A property is an attribute of the object. The object concept is the sum of all the properties. The object owns or contains the property.

Actions

Some actions of this language are:
Read to input data, or to get next.
Write to output data, or to put next.
Add
Subtract
Multiply
Divide
. . .
I note at the beginning, that all actions that are developed here will be done in computers. There is a computer for each action. The proper type of computer will be developed before we get to Hal Algebra. It must read and write as well as do functions.

Actions with One Object

With one object you are restricted in action. With number symbols, you can only count. Counting is the most primitive action you can do. It is Count = Count + 1 .

Or uncount = uncount - 1 .

Actions with Two Objects

With two objects one can do arithmetic. All actions are possible. What is lacking is anyplace to store the result of the arithmetic. You can Read but not Write.

Actions with Three Objects

With three objects and an = sign, you can make an algebra. Now you can Read and Write.

A system is a set of coordinated objects and actions.
A math system is a set of coordinated math objects and math actions.

Some math objects are: Number, signs, symbols

Some properties of math objects are: Value, name, position, size, and base.

A number can be a connected set of digits.
N = {Dn, Dn-1,... D0}   123

Digits are an ordered set of symbols.
D = {S1, S2, ... Sn}   in 123, $1*10^2$ is digit 1, $2*10^1$ is digit 2 and $3*10^0$ is digit 3. Base is 10.

1.   Base is the count of symbols. Base is a counting number. {0 to n}

2. Digit has b names and values of 0 to b-1. All digits cycle. Here is a b = 3 digit:

name,	count,	value
0	1	1
1	2	2
2	3	3
0	4	1
1	5	2
2	6	3

3. Numbers are made from digits or are a digit. With multiple digits of the same base, numbers are polynomials in b.
Here is a b = 3, two digit number: d*b + d

d1d2	d*b	d	value
00	0	0	0
01	0	1	1
02	0	2	2
10	3	0	3
11	3	1	4
12	3	2	5
20	6	0	6
21	6	1	7
22	6	2	8

D1*b + d2 = value

This is also a b = 9 single digit. Any number can be converted into a single digit by letting
$b[new] = b[old]^{digits}$ 9 = 3^2

Real numbers are two digit numbers, I and F. The digits I and F are connected but they can be of any and different base. Here base will be attached to the name.

Ib and Fc are names with base b and c.

I2	F3	count	value	I3	F2	count	value
0	0	1	0.0	0	0	1	0.0
0	1	2	0.1	1	1	2	1.1
0	2	3	0.2	2	2	3	2.2
1	0	4	1.0	3	0	4	3.0
1	1	5	1.1	0	1	5	0.1
1	2	6	1.2	1	2	6	1.2

End of first cycle.

Note that system base is b*c. There is no decimal point in nature. Decimal placement can be anywhere in the number without effecting the count and it works as a place marker.

Multiplication and division move the decimal point while addition and subtraction do not, except when carry/borrow occurs.

To read a real number, read I then read F. To write, write I then write F. (This is a rule, not a requirement.) Note that any number systems states can be counted and thus related to a counting number.

The real number from 0.0 to 0.1. The count

0.00	0
0.01	1
0.02	2
0.03	3
0.04	4
0.05	5
0.06	6
0.07	7
0.08	8
0.09	9
0.10	10

4. Variables are containers of numbers. Variables have a name and content or value. If A is a variable name then the statement A = 2 gives it the value 2.

5. HalStreams are variables that flow in time. HalStreams have names and their content is the flow of variables. To make variables flow, one eliminates coordinate systems and adds time. Variables can now be

used whole or a datum can be reached by indexing into the stream. If S is a HalStream, then S[j] = S[j-1] always. How this flow occurs will be shown in the actions of Book 3.

Some actions are: +, -, etc. and depend on the objects they work on. With actions we need definitions. Here is a procedural definition of an equation. O, L and R are internal variables.

```
Plus
 Read L          ;The variable L gets set.
 Read R          ;we stop at two inputs
 O = L + R       ;do the action, here
addition.
 Write O         ;emit the answer
 GoTo Plus       ;this makes a HalStream.
STOP is a static.
```

For equations, we need one of these for each action. This is why functions were invented, so you could talk about any equation. Here is a two input function:

```
F
  Read L
  Read R          ;two inputs
  Do O = F(L,R)   ;any computer program here
  Write O         ;O is defined by and caused
by F(L,R)
  GoTo F          ;make a HalStream. Use STOP
for static.
```

Note that equations can only do their
specified action. Functions can be anything
you can program that relates L and R to O.
This shows how time added to variables
produces a HalStream. The cycle time of
these procedures is c, which is the sum of
the statement times. That means that O is
emitted every $1/c$ seconds or a new datum is
there in c seconds. A HalStream is its own
coordinate system, x,t.

6. Equations contain some action and
variables or HalStreams. $Z = X + Y$ is an
equation with two inputs. Equations are
input, output devices and the number of
inputs and the action are its attributes.
Also X + Y defines Z. Since X and Y are not
defined, they are sensors or axioms.

7. Equation HalTrees are equations arranged in a HalTree form of n inputs and one output. HalTrees are equations in time and space. HalTree dimensions are time and distance. The n inputs form a column of a time. The next column is next time and it is n-1 (full HalTree) or n/2 (sparse HalTree) units high. S[x,t] is an equation location. Here is a 4 input full HalTree to illustrate.
S[1,1]
S[2,1] + S[2,2]
S[3,1] + S[3,2] + S[3,3]
S[4,1] + S[4,2] + S[4,3] + S[4,4]

HalTrees can be multi dimensional, with t being one dimension. (S[x,y,z,t] for example.) The 4 input HalTree is a slice through a higher dimension HalTree. The dimensions are not locations of data but are location of the equations and since equations are physical nodes, they must be in space-time.

8. Functions are the variables of
equations. Z = f(X,Y).
Functions are also input, output devices
and number of inputs is the main attribute.
Action is left as an open variable. The
value of a function is the value of its
equation.

9. Function HalTrees are like equation
HalTrees with the equations not defined.
Here is the 4 input HalTree with the
function f with S[x,t].
S[1,1]
S[2,1] f S[2,2]
S[3,1] f S[3,2] f S[3,3]
S[4,1] f S[4,2] f S[4,3] f S[4,4]

Now a table can be made to show these relationships.

no.	object	contains + adds	example
1.	b		n
2.	digit	b, value	3
3.	number	digits, polynomials	123
4.	variable	number, name	name
5.	equation	stream, action	
6.	function	equation,	S = f(x)
7.	HalTree	Function, multi inputs	

HalTree(function(equation(variables(number(
digits(b))))))

I write the symbol 8. Is that a
number?

Book 2
Numbers and Number Wheels

Human mathematics arose from the needs of accounting and measurement. Counting and comparing of static quantities is inherent and automatic. I see that Joe has more junk cars in his yard than Jim has in his yard but fewer than I have in my yard. I do not have to count them. I can see everything at a glance. Humans can see small numbers of objects. None, one, two, three and a lot are sufficient number names for most discourse. Crows and cats can distinguish three things at once. Animals without words see "this thing," "this thing and that thing" and "this thing and that thing and that other thing" at the same time. We humans say "one, two and three things" since we have the words to say it.

The language of mathematics grew out of spoken language. It is a made language, like a computer program. There can be any number of mathematical language systems. There is nothing special about the ones we have and there is much that is suspect. There is nothing special about this system. It is merely a change of viewpoint and focus. Yet we know the universe is a calculator, strictly obeying the rules of nature. It is those rules we wish to tap. Human mathematics we leave for play.

All languages consist of coordinated actions and objects and can be any set of coordinated actions and objects. The purpose of any language is to compose and decompose sentences in that language. The original inventors who wrote the first sentences in the language of mathematics were really creating programs for computers (Human computers at that time.) So the language of mathematics grew in all directions, like any computer program written by multiple authors without insight into what they were really doing.

The invention of the electromechanical computer negated all previous mathematical programs. This has not bothered mathematicians any more than Copernicus's elimination of Heaven and Hell bothered Christians. Number loses its magic and infinities and gains its place as elemental.

Pythagoras was almost right in his vision of the world built of stacks of numbers. The world is not built from numbers, but from the actions of arithmetic. Matter holds together because of the necessity of atoms to balance their charge accounts. Addition and subtraction start here. Positive and negative charges must balance. $8 = X - Y$ is the arithmetic of molecules. Numbers are not things. They do not stack. However, nature is bound by the rules of arithmetic.

Atoms buy and sell electrons and the books of charge must always balance. The constant integration of gravity is the other organizing arithmetic, constantly computing the curvature of space. We cannot see this arithmetic or its numbers, but we can abstract the actions and see some rules.

A protein molecule looks like a pattern of electrical charges. A pattern of charges can be seen as a number. An assemblage of protein molecules arranges itself to minimize charge differences. In the abstract, they compute charge balance. The actual mechanism of computation we leave to Biology. There are shape changes and changes in orientation as well as changes in time. Every molecule attracts and repels neighboring molecules, each influencing the orientation of the other. Charges move molecules. Molecules move charges. There is no conventional mathematical solution of this kind of computation. This is why data stream mathematics was invented. Interest is shifted from solving the molecule state problem, to building a machine to act on the molecular data itself.

Numbers are adjectives, attributes of collections of things and do not exist except as an attribute of something. For the last 3000 years, there has been confusion about this. Numbers and symbols of numbers are not the same thing. Most of the literature that mentions number really means number symbols. If number symbols were numbers, H_2O would wet the paper or the computer screen here. We need a device to clarify the difference between number and the symbol of the number.

Talking about numbers requires a viewpoint or device to give body to an abstract notion. The ancients used figures with lines and points. They used a line and points to represent numbers. Here is a modern number line for integers:

...-5 -4 -3 -2 -1 0 1 2 3 4 5...

The ancients thought numbers were continuous because the continuous line they were drawn on made them appear that way. Early mathematicians deduced that they could add one to any number they could imagine, so there could be an infinite count. They could not conceive of changing the base of the numbers on the number line, so all fractional and real numbers were squeezed in between the natural numbers. This created infinitesimals. Numbers and symbols of numbers have always been scrambled and interchanged. We need a tutorial device to separate symbol and number for us. In the process, we will see that there is only one number system. Everything else is a symbol.

When one is talking about number, one must allow space for symbols. We work with number symbols. Symbols of numbers are not numbers, yet, under every set of symbols lurks the counting numbers. We can always count the symbols. The symbol count is the primary property of the number system.

The first thing to do is to roll up the number line and connect it into a circle. You know how to do that when you make circular functions. From the circle, make a wheel.

A list of number symbols of any size can be cut, pasted and looped onto the wheel. Make a mechanism to fit on the axle of this wheel to click from one number symbol to the next. There is a pointer or window to indicate the active symbol. Some early electronic instruments had thumb wheels to input numbers into the processing unit. These are an exact analogy to number wheels. They click from one number to the next exactly and have a window so you can only see the one digit. I have one on my desk. You can do arithmetic with it in a primitive way, (by imagining another wheel).

The wheel is created and a clicker is added. The wheel can be more like a drum with space on the circumference for the symbols. Put a fixed window so only one number space can be seen. The window is attached to the fixed part of the click mechanism. We are looking at the window showing 8 or Present of I or .8 or anything else this click could show.

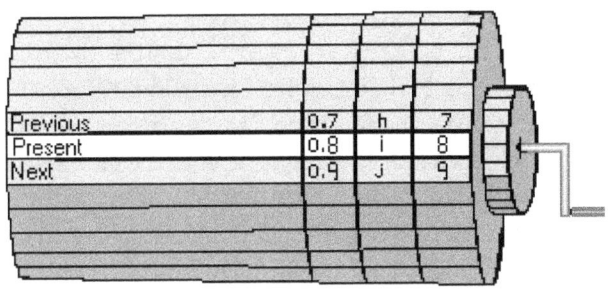

Figure 1 Number Wheel

Count the symbols on the wheel. Add one for the zero symbol. That count will always equal the number of clicks on the wheel and is **b**. To construct number symbols on the wheel, start anywhere and draw the symbol **0** on the wheel. Click forward one click and paint a **1**. Click once more and paint a **2**. Continue painting successor symbols until the 0 symbol occurs in the window. The number wheel or digit wheel is done. The **b** clicks are the "base" of the number, and all clicks on the wheel are base **b** arithmetic. **b** is also the count of symbols, including a zero. Note that no **b** was associated with the number line. The base of a wheel or any bit limited number must be specified. We lose infinitesimals but we gain **b**. When we made the wheel we lost infinity. Now we can only count to **b**.

   b is a count of symbols, therefore it is a positive whole number. Since **b** can be any value and since there is no symbol for **b** in the set of symbols in **b**, what happens when **b** is 0 or 1?

b = 0 means there are no symbols and no clicks. We can take it to mean that all **b** = 0 numbers are 0.
Count is 0 0 0 0…

b = 1 means there are still no symbols and no clicks. We can take that to mean that all b = 1 numbers are 1.
Count is 1 1 1 1 1 …

b = 2 has the symbols 0 and 1 and has 2 clicks to cycle from 0 to 0. Count is 0 1 0 1 0 1 … This is binary. Now we can do logic.

b = 3 has the symbols 0, 1 and 2 and has 3 clicks. Count is 0 1 2 0 1 2 … We can also count -1 0 1 -1 0 1… making the first integer. (Because there is room to assign a sign bit.) ...
b = 7 has the symbols Sunday, Monday, Tuesday, Wednesday, Thursday, Friday and Saturday. It has seven clicks. Count is Sunday Monday Tuesday ... This is a days of the week wheel.
• •
b = 10 has the symbols 0 1 2 3 4 5 6 7 8 9 and has 10 clicks. Count is 0 1 2 3 4 5 6 7 8 9 0… This is good old base-10 stuff. Now we can write number polynomials directly.
...

B = 360 is the degrees in a circle base.
...
b = n has the symbols 0 1 2 … n-1 and has n clicks. Count is 0 1 2 3 … n-1 0 1 2 … n can be a large number but it cannot be larger than the number of symbols we can express. Ultimately it relates to the number of switches or wires we can make.

Here is a list of wires, bases, symbols and how they count.

Wires	b base	Symbols	counting	Comments
0	0	0	0000000000000000	All zero
0	1	1	1111111111111111	All ones
1	2	0-1	0101010101010101	Binary
2	3	0-2	0120120120120120	Integer
3	4	0-3	0123012301230123	
3	5	0-4	0123401234501234	
3	6	0-5	0123450123450123	
3	7	0-6	0123456012345601	
3	8	0-7	0123456701234567	
4	9	0-8	0123456780123456	
4	10	0-9	0123456789012345	Decimal
4	11	0-10	0123456789A01234	
4	12	0-11	0123456789AB0123	
4	13	0-12	0123456789ABC012	
4	14	0-13	0123456789ABCD01	
4	15	0-14	0123456789ABCDE	
4	16	0-15	0123456789ABCDEF	Hex

b controls the number of symbols and the cycle and implies the number of bits or wires.

b is the number of clicks on the wheel, the base of the number and the number of states of the state machine we could make.

Each number symbol requires space for its picture. This means the wheel will have to have size to carry all the pictures. If the number picture size is **h** units high, then the circumference of the wheel is **b*h**. **b** is the base of the wheel. The digit system base is simply the number of symbols, assuming one symbol for each click. This includes the zero symbol. If there is a circumference, then there is a diameter of **b*h/3.14159** (assuming flat space.) If we have 10 symbols, including a zero, our base is 10, and if the number size is 1 inch, then the wheel diameter is 3.18 inches. (Make a wheel like this out of a flat piece of cardboard. Cut out a circle and write symbols around the circumference. Stick a pin through the center, so you can spin it.)

The unit circle and circular functions are the connection to math.

To go back to the ancients, make **h = 0**, and **b = infinity** and you have their number line rolled up into a circle. The circle has a diameter of 0 so don't look for it. Pythagoras would love this. Space is made up of number circles of 0 diameter, he would say. Expand space and get a number wheel. Rotate the wheel and do arithmetic. Wheels make a lot more sense than vibrating strings. Of course, they may be the same thing.

A nice wheel is a base-10 digit with a
picture size we can see easily.

The wheel clicker has exactly **b** clicks to
cycle the wheel. For example, if our wheel
is sitting on the symbol 8:

```
Click     1 2 3 4 5 6 7 8 9 10 11 12 …
Symbol 8 9 0 1 2 3 4 5 6 7  8  9  0 …
           |_____ A cycle _____|
        <-------- Wheel Direction ------->
```

Click could be the output of an INode.
Click = i(1)

0	1	2	3	4	5	6	7
0	.1	.2	.3	.4	.5	.6	.7
0	a	b	c	d	e	f	g
0	2	4	6	8	10	12	14
0	2	4	8	16	32	64	128

Note that we can expand the wheel into a drum and write many symbols side by side by putting them in the same base.

<------ Wheel Direction ------->

You can write every number symbol set that fits in a base-8 system, including all old symbols like Roman numerals and any new Green Alien symbols to come and, underneath it all, is our counting set of number clicks. Galileo used this side-by-side device to prove there are as many even numbers as there are counting numbers. What he really proved was that there were the same number of symbol space required to display them. There is no number line. There are no infinities in nature. Infinities and zero diameter number wheels are mind puns. Behind everything is N the natural number set. That is irreducible.

Clicks can be generalized to any "get next" and "get previous" operations. The symbol on the wheel does not need to represent a number. If "a" is in the window of the wheel and we click upwards a click, the next picture is the result of the addition. On the other hand, every "get next" and "get previous" operation can be reduced to wheel clicks.

Symbols are visible. Clicks are counted and are abstract. The 2 symbol I wrote at the beginning of this can now be seen as a mark, a symbol. The wheel separates symbol and number for us. The symbol is painted on the wheel. The number is the abstract click count. In a computer the click would be a fetch mechanism, and the wheel would be the computer memory. A rotating or oscillating wheel presents a moving data stream at the window.

Ground, Power and Electric Numbers

b = 0 can also be interpreted as electrical ground. Ground value is supposed to be zero at all times.

b = 1 can be seen as electrical power supply voltage. This means electrical values are between base 0 and base 1 and are not a number.

To make digits real so they can be manipulated and seen, we make an electrical analog. Digits are carried on wires now. The symbols are off and on and b = 2 for each wire.

Start with one wire. Call it P. Make a table.

Wires	Names	Capacity or usage
1	P	None
2	P,G	Power and ground
3	P,G,S	Pwr, Gnd, signal
4	P,G,-P,S	Pwr, Gnd, -Pwr,Signal

1 wire   Nothing can be done with one wire. No flow can be made.
2 wires   Connect two wires to a battery. With a good power supply, P - G = constant.

3 wires   P – G = K. S – G = variable. Now
we can define the number S as gradations of
P – G. If P – G = 9 then we can make
numbers from 0 to 9, or 0 1 2 3 4 5 6 7 8 9
as gradation values. We have sort of a
positive digit here. Computation can be
done as long as we stick to all positive or
all negative numbers. To fix that we need:

4 wires. Now numbers are graded by the
distance between P and –P. Still V = S – G
but now V can become negative when S less
than G. With this V we can do almost any
computation, including subtraction. Since
we can do almost any computation, why did
these analog numbers not take over?

      The problem arises when P – G is not
constant. If that happens, our number
gradations go to pot. If the power supply
is noisy, so is our number. This is not
just a noise problem. When the P – G
changes, our number base changes.
      The great advantage that analog
numbers have is that they already flow
because the defining electricity is a flow
of electrons. Electrons flowing through a
resistance create a voltage difference
across the resister. That difference is our
number. We have no number here without
electron flow.

To fix the P - G voltage change fault, we make the number definition worse and we go back to three wires.

The number is derived from S , but we define a new number B this way:
If S is near P then B = 1. If S is near G then B = 0. For example: Let P = 5 and G = 0. B = 1 I S > 3 else B = 0.
There are just two gradations between P and G. That seems worse, but now P and or G can vary a lot before the number will change from 1 to 0 or vice versa. We make B robust. It becomes a base 2 digit. With base 2 digits, one can do Boolean and switching algebra. There are two symbols or conditions for base 2, off and on or 0 and 1. To make a general digit, use multiple binary digits. A four digit binary number is:
$BN = d*2^3 + d*2^2 + d*2 + d$ where d is 0 or 1.

A general number expressed with digits is:
$GN = d*b^3 + d*b^2 + d*b + d$ where d is < b.

Multiple Wheel Numbers

We are accustomed to having multiple number wheels. We count 0 1 2 3 4 5 6 7 8 9 **10**… We start a new digit wheel with hardly any thought.

When you count in any number base and run out of symbols you have three choices for continuing. You can just let it roll over or you can create a new symbol to expand b or you can create a new wheel and a mechanism to hook them together. To count, click the digit wheel. Note that you must be at, or set, a starting place on the wheel. A count is a get next click. When you get back to the zero symbol on the wheel you need help to "carry" the overflow.

Hire another counter to work the next digit wheel. The instructions to that counter are: "when you see a zero symbol appear and disappear then click your wheel one time." No matter how many digit wheels you have beside the first, this is the instruction required. You, on the first wheel, can count by ones, by twos and so on, but the other wheels can only count by one when the wheel below it cycles. Multiple wheel, or multiple digit numbers can all be expressed as polynomials in b as follows:

Positional notation writes every number as a sum of powers of b. An 7 digit binary number is:

BNumber = D_6*2^6 + D_5*2^5 + D_4*2^4 + D_3*2^3 + D_2*2^2 + D_1*2^1 + D_0*2^0 Where D_x is {0,1} b = 2
2^6 = 2*2*2*2*2*2

A 3 digit decimal number is:
DNumber = D_2*10^2 + D_1*10^1 + D_0*10^0 = $D_2D_1D_0$
Where D_x is {0 through 9} b = 10

An N digit base b number is:
EveryNumber = D_n*b^n. . . D_3*b^3 + D_2*b^2 + D_1*b^1 + D_0*b^0
Where D_x is one of b symbols, including 0.
b = b

There is no requirement that all digits have the same base. In fact the above equivalence breaks down if any digit has a different base. The sum no longer is the base 10 value of the multi digit number.

It takes time to click the wheel. Let
that time be represented by the symbol c.
Assume every wheel in this system takes the
same time to click. I do not have to assume
that, but it makes the following exposition
messy if I don't.)
The time required to cycle the first
wheel is b*c where b is the number of
symbols on the wheel and c is the cycle
time per symbol. (To represent a number,
every wheel in the system has the same
number of symbols.)
t0 = b*c   ; is the cycle time of the first
wheel
t1 = b*t0 ; The cycle time of the second
wheel
t2 = b*t1 ; and so on.

If you sum the times to get a total time,
you get:
ts = t2 + t1 + t0   ; the
sum
ts = b*t1 + t1 + t0   ; t2 =
b*t1
ts = b*(b*t0) + b*t0 + t0   ; t1 =
b*t0
ts = b*(b*(b*c) + b*(b*c) + b*c   ; t0 =
b*c
ts = c*b^3 + c*b^2 + c*b ; which is the very
picture of the polynomial number based on
the time required to make it. Powers are
related to counts of digit wheel rollovers.

A base 10, 3 digit number is:

$$b10 = d2 * 10^2 + d1 * 10 + d0$$

Now when you look at a number like 1998 you can see it as a set of four wheels. Wheel 0 is 80% exhausted. Wheels 1 and 2 are 90% done and wheel 3 is 10% done if it is a base 10 number.

Just looking at a number symbol like 1998 does not tell you what its base is. You know it must be at least a base 10 number because it has a 9 in it. But it could very well be a base 11 or above.

By manipulating symbols and base, you can always get rid of multiple digit numbers. 123 is a 3 digit number in some base greater than 3. By treating 123 as a single symbol instead of as three digits we can make it into a single digit. It is a single digit number in some base greater than 123. You can make a single digit base 1000 number with 999 symbols and zero. 123 then is a single symbol, a single digit, in this system.

Any number can be made into a single digit by this process. In general the new base is:

$b = oldbase^{digits}$. Since any number can be converted to one digit that means that every number must cycle.

Real Wheels

A wheel of real numbers illustrates the fractal nature of all numbers. The number base, as always, is the count of all the symbols. Let r = the number of real symbols before the decimal point and f = the number of symbols past the decimal point. If r,f equals 10,10, then the clicks are scaled to .01 each click and the wheel is the same as a base 110 wheel.

[b = f+1+(f+1)*r] ;is how to compute base from r,f

There are no decimal points in nature. We can always put the counting numbers on the same wheel to keep it honest. For illustration, here is a wheel of r.f = 2.2 where b = 3+3*2 = 9 : which means 9 counts from 0.0 to 2.2 with counting numbers.

```
. 0.0 0.1 0.2 1.0 1.1 1.2 2.0 2.1 2.2 .
Real numbers
.   0   1   2   3   4   5   6   7   8  .
Clicks. b = 9
```

To compute a proper base, you should make the maximum value of real number to work from. Treat the r part and the f part as separate wheels. In the example above, start with 9.9 and compute that base. We

have 10 + 10*9 = 100 which is a proper
base. We count 0.0 0.1 0.2 ... 9.8 9.9 real
                              0   1   2 ... 98
99 b = 100
    For base 8 digits we have 7.7 whose
base count is:
8 + 8*7 = 23 meaning we need 23 decimal
counts to list from 0.0 to 7.7 with base 8
digits.
7.77 is 80 + 80*7 = 640
7.777 is 800 + 800*7 = 6400
See what fractions cost in computing power.

Irrational numbers

Irrational numbers arise from division, so they do not appear in the Hal algebra system. However, we still might have to pay attention to them.

On the surface it would appear that irrational numbers would require that the base of the number would be infinite. Indeed, that is so. In practice, we make a wheel of the precision required, thus avoiding making infinite symbols. The wheel for Pi, for example could have a base of 1000000. This would accommodate six digits. 3.14159 could be coded as 314159 which is contained within 1000000. Those symbols we can write easily.

If we need an irrational constant, we can provide it, by making a system of real arithmetic to replace the integer arithmetic. This adds complication and I do not think nature works this way.

Clicking Around the Number Wheel

Given a number wheel, we can count or add by clicking in one direction. It is possible to click right around the wheel. I mean that we can do more clicks than there are numbers on the wheel. What happens when we do? It is important to know because all numbers in nature are bit limited and must at times roll over.

There are two parameters to manipulate. There is the number base of the wheel, b and the integration number, k. Compute $X = i(k)$ on the base N wheel. On the wheel and in nature this calculation is automatic. When you do this on a calculator, thus simulating the wheel, you must use a procedure. Start with $X = 0$. Add k to X until X is $> b$. When this is so, subtract b from X, write down the number in the calculator and continue.

Each time you do the subtraction of b from X means the wheel has revolved past 0 and $X - b$ is the landing spot that starts the next integration. In the examples below, record only the landing spots to determine the cycles. We are not interested in the numbers in between. The numbers that are recorded are made bold.

Generate numbers until the wheel reaches 0 again.

Here is the integration of a base 8 number
with the rollover points made bold.

How computed

i(1)=	**0**	1	2	3	4	5	6	7	**0**
i(2)=	**0**	2	4	6	**0**	2	4	6	**0**
i(3)=	**0**	3	6	9	**2**	5	8	**1**	0
i(4)=	**0**	4	**0**	4	**0**	4	**0**	4	**0**
i(5)=	**0**	5	**2**	7	**4**	**1**	6	3	**0**
i(6)=	**0**	6	**4**	**2**	0	6	**4**	**2**	**0**
i(7)=	**0**	7	**6**	**5**	**4**	**3**	**2**	**1**	**0**

Here are the rollovers

0	0						
0	0	0					
0	2	1	0				
0	0	0	0	0			
0	2	4	1	3	0		
0	4	2	0	4	2	0	
0	6	5	4	3	2	1	0

Let Sunday = 0, Monday = 1, Tuesday = 2,
Wednesday = 3, Thursday = 4, Friday = 5 and
Saturday = 6 and we have a base 7 wheel. We
roll that to start.

----

b = 7   **x = i(k)-c**   **Start at x = 0, c = 0**

k	0	1	2	3	4	5	6	0	Revolutions
1	0	1	2	3	4	5	6	0	1
2	0	2	4	6	1	3	5	0	2
3	0	3	6	2	5	1	4	0	3
4	0	4	1	5	2	6	3	0	4
5	0	5	3	1	6	4	2	0	5
6	0	6	5	4	3	2	1	0	6
7	0	0	0	0	0	0	0	0	7

Figure 2 Base 7 Wheel Spins

This is base 10 wheel cycle.

b = 10   **x = i(k)-c**   **Start at x = 0, c = 0**

k	0	1	2	3	4	5	6	7	8	9	10	
1	0	1	2	3	4	5	6	7	8	9	0	1
2	0	2	4	6	8	0	2	4	6	8	0	2
3	0	3	6	9	2	5	8	1	4	7	0	3
4	0	4	8	2	6	0	4	8	2	6	0	4
5	0	5	0	5	0	5	0	5	0	5	0	5
6	0	6	2	8	4	0	6	2	8	4	0	6
7	0	7	4	1	8	5	2	9	6	3	0	7
8	0	8	6	4	2	0	8	6	4	2	0	8
9	0	9	8	7	6	5	4	3	2	1	0	9
10	0	0	0	0	0	0	0	0	0	0	0	10
11	0	1	2	3	4	5	6	7	8	9	0	11
12	0	2	4	6	8	0	2	4	6	8	0	12
13	0	3	6	9	2	5	8	1	4	7	0	13

k > b starts over at another scale. Each count requires one more revolution on this scale. Everything repeats at this new scale and this can continue with larger k until k runs out of bits.

You will notice three rather
astonishing things:
1. All wheels cycle back to the starting
point.
2. Odd k numbers require exactly k turns of
the wheel to cycle regardless of number
base except when k = b/2.
3. When k is greater than b, we start
counting again, starting another digit.
This is another cycle, on another scale.
Wheels are fractal. Numbers are fractal.
This is what really causes chaos.

Numbers are made of digits. The digit
is the foundation of number.
A digit is made of symbols or signs.
In a sense, a digit is the list of symbols
or signs. A digit also has position
information. The position in a number is
important. A digit, at a given time, shows
a value, (one of its symbols.)
The number of symbols or signs is
called the base of the digit. A digit is
always multiplied by its base raised to its
position. This is a virtual multiplication
that does not show on the digit.
Here is a number: 1234
It has four digits showing 1 2 3 and 4.
Base is not shown in the number, so it
could be anything greater than 4. Let b be

the symbol for base, D be the symbol for digit value and P be the symbol for digit position.
Then D*b^P defines the digit.

ABCD

$A*b^3 + B*b^2 + C*b + D = S$

A multi digit number is a polynomial in powers of b.

Given two rules, any number can be made into a single digit.

1. Reuse number symbols. …9 10 11… where 10 and 11 here are considered one symbol each.
2. Let base, b = oldbase^digits.

Taking 1234 and making it a single digit is to replace the base 10 with 10^4. Now we can count directly with no carry to 9999. That is not $9*10^3 + 9*10^2 + 9*10 + 9$, it is the single digit 9999 which will go to 0 when you add 1.

Obviously we need some better nomenclature here.

Digit size is limited by the number of symbols. Is there a limit on digit polynomials? If not we are back to infinities.

D1	D1
D1 D2	$D2*b + D1$
D1 D2 D3	$D3*b^2 + D2*b + D1$
D1 D2 D3 … Dn	$Dn*b^n +… + D2*b + D1$

The subscript and power n has no relationship with b and can grow as long as we can add digits. But n is a digit and must have a base. This puts a limit on the size of the polynomial.

Book 3
The computing machine as a math object.

The computer

The idea of a computer doing math was simultaneous with the computer as a mathematical object. The notion that a computer program could do calculation became that of studying the computer as a mathematical device. Unfortunately, the early mathematicians who did this knew very little about computers, so none of them got the architecture right. Neither Turing nor Chaitin could know about microcontrollers at that time. The idea of computation at that time was of computing the answer to some problem and then stopping. The mathematical idea was whether one could predict if the computer would stop or finish the computation. (I am going to reverse that so that, when my machines stop, it is because of an error in programming or the world has come to an end.)

A computer is two machines, a fetch machine and a decode machine. A linear array of memory cells and registers ties them together. The computer of interest now is the microcontroller that has input, output ports as well as registers. It also has a program memory array that can store a program outside the register array. The microcontroller (MC) has all the elements of any computer.

The MC is also two machines. There is a mechanism to fetch program code from memory. There is a decoder to decipher these codes, and there is an execute engine to execute the codes.
The sequence is:
    --- Fetch -> Decode -> Execute -> Fetch ---
Note that this is a simple loop. This is all a computer does and can do. Decode and Execute are always tied together as a unit.

Decode parses the code provided by Fetch. For every operation code, there must be a response. Illegal op codes will suspend operation and will set the fetch address to an error trap. Legal codes, memory address and so on are sent to the Execute unit.
The execute unit performs the actions of the computer. This may involve more memory access, and may change the program counter address.

Computer hardware is congealed
software. It is a design choice of when
quit hardware and resume software. From
above, the user cannot distinguish between
hardware and software. Therefore, we can
use software descriptions of any hardware.
We can simulate or we can build. This of
course is what makes HalNodes work.

Px is an input or output port.
C is program counter. Cwc is width of
program counter. Program counter must be
able to address all memory.
Rx is register. Rw is width of register.
F is flag register.
(in a regular machine, w is a constant,
i.e. all registers are the same size.)

If a processor is to do anything, it
must make a change in memory. We should do
at least these things:

1. IF F then P = M command: 00 ;Jump on condition
2. Read P to R   command: 01 ;R1 = P1
3. Write R to P   command: 10 ;P1 = R1
4. copy R1 to R2   command: 11 ;R2 = R1

requiring 2 bits. So Dwd = 2

With this 2 bit processor, we can move
data, and we can change data, but we cannot
do any arithmetic until we have an Adder.
And just to command the arithmetic, we need

more commands; that means we will need more bits.

Adding 1 bit doubles the number of commands each time. With 3 bits, we can have 8 commands.

Bits	Commands
1	2
2	4
3	8
4	16
5	32
6	64
7	128
8	256
-	-
32	4294967296

or Commands = 2^{Bits}

The Turing machine is a state machine, which is not a computer. Let the Turing machine be TM. Then
TM = {tape, read, write, state}

The register machine gets closer to the computer, but the actions are wrong because there is no input or output. Let the register machine be RM. Then
RM = {registers, operations on registers}

The microcontroller is a better math machine, since it maps to functions with input, output ports and has memory for data and program like a real computer. With ports, MCs can talk to each other, thus forming tree networks. Let the microcontroller be MC. Then
MC = {ports, mem, code}
MC.ports = {I, L, R, Out}
MC.mem = {I, L, R, M, Out}}
MC.code = {labels, statements, comments}

MC.code.labels = {Yes, No, Done, Loop, BEGIN, END}
MC.code.statements = {Read, Write, GOTO, IF, =, +, -}
MC.code.comments = {anything after this symbol ;}

BEGIN Example
LABEL1
  STATEMENT1 ; COMMENT
  STATEMENT2
LABEL2
  . . .
  STATEMENTN
LABELN
END Example

The MC I will use to make a math object must have at least three ports and corresponding registers. We can name them L, R and O. There are a finite number of registers that can be devoted to programs. Mathematical variables are exactly registers and will not be distinguished from now on. The necessary MC operations for a mathematical computer object are:

```
READ       ;Reads data at a port and sends
it to a variable
WRITE      ;Takes a variable and writes it
to a port
IF Cond    ;Does true or false for Cond
(Goes to Yes or No)
GOTO       ;Changes fetch address to named
label
Arithmetic ;At least +,-, =
```

We establish as well, all the arithmetic and logic operations and data movement from register to register. All this requires only 8 bits but we can imagine more. A program is an ordered list of operation codes. Here is a program for any two input function:

```
BEGIN Node
Loop              ;This is an address name
  Read L          ;Read a port L to L
  Read R          ;two inputs is all we ever
need
  Out = F(L,R)    ;Any function code goes here
  Write Out       ;Write Out to port O
  GOTO Loop       ;do this forever. Make a
data stream
END Node          ;this is instruction to the
assembler.
```

In a real MC, we would have to tell it which ports are input and which are output as well as other housekeeping things. Mathematically, this is not important.

The usefulness of the MC as a math object is that we can exactly define complex functions and see how they work with this tool. For example, most math functions now are:

```
BEGIN Function
 Read L          ;get L ,L is both port and
variable here
 Read R          ;get R
 Do O = F(L,R)   ;do any function
 Write O         ;display it to O
 STOP            ;and stop
END Function
```

Variables and ports are different objects, so they can have the same name. The alternate is to add a TO code. Then the code for ports = {A,B,C} and variables = {L,R,Out} would be:

```
BEGIN Function
Loop
  Read port A to L
  Read port B to R
  Do F(L,R) to Out
  WRITE Out to C
  Go to Loop
End Function
```

You could then have Read Port L to L made explicit.

"To" is implied in every statement, whether we say it or not.

Note the difference in a static function and a function that makes a stream of data. This difference is a change in one statement that is outside the function, so we can separate data action from the function, meaning any function can be a stream of data.

We can leave off the BEGIN and END so
our general function code for data streams
is:

```
G Function     ;name of node
Loop           ;label
 Read L        ;Get L
 Read R        ;Get R
 O = F(L,R)    ;Do any Function
 Write O       ;Put O
 GOTO Loop     ;do it forever
```

With this verbose notation we can
examine the form of functions, and later,
of arithmetic.

A single input function is:

```
Single
Loop
 Read I
 O = F(I)
 Write O
 GOTO Loop
```

I, L, R, and O are the internal names of registers in the hardware nodes. This means they can be given any name outside. So your code might read:

```
MyNode
Loop
 Read MyInput
 MyOutput = F(MyInput)
 Write MyOutput
 GOTO Loop
```

Functions With Memory

Functions as machines can be simple, as above, or can have memory as will be shown.

```
MemFunction
 M = 0            ;set memory initial
condition
Loop
 Read I           ;single input
 Out = F(I,M)     ;double function
 Write Out        ;single output
 M = I            ;refresh memory
 GOTO Loop        ;do it forever
```

For example, here are difference and sum functions.

```
Difference        ;differential Out = d(I)
 M = 0            ;initial M
Loop
 Read I           ;new data
 Out = I - M      ;difference of new and old
 Write Out        ;write it
 M = I            ;replace old with new
 GOTO Loop
```

```
Sum                   ;integral Out = i(I)
 M = 0
Loop
 Read I
 Out = I + M    ;sum of old and new
 Write Out
 M = I
 GOTO Loop
```

The only difference in the two procedures
is - -> + and the names. Note that:
Difference(Sum(x)) = x.

Conditional Arithmetic

We can now do functions of conditional objects like GT and LT. The conditions are: L > R and L < R. Now the IF statement is required. We write IF L > R, GOTO Yes, GOTO No where Yes and No are labels for coded that does the true or false of the condition. This is just like the spread sheet form IF(condition, true, false). This fits the MC better than the usual IF cond Then Yes else No which can get entangled in else ifs.

```
Conditional
Loop
 Read L
 Read R
 IF L cond R, GOTO Yes, GOTO No
Yes
 Out = L   ;or whatever is true
 GOTO Done
No
 Out = R   ;or whatever is false
Done
 Write Out
 GOTO Loop
```

Out = L condition R is how this would be written
Z = L LT R is an example.

Math Machines

The machine as a math object is one thing. Now I make math into machines.

```
Math = {objects, actions}
objects = {variables, sets,...} ;How
```
many objects are there?
```
actions = {+, -, *, / ...} ; How many
```
actions are there?

Variables, in this system are inputs, output, registers or memory.

Sets cannot be mechanized as is because there is no way to get data into and out of a set. There are no read, write operations on sets although we could easily make them. However, data streams variables are open sets, so we do not need both.

Functions are machines that have inputs, domain, and one output, range. There is a program in the function machine but it can be any program.

Equations are machines with inputs, an output and a fixed program. A = B + C can be written A = +(B,C) and A is the output of the + machine with inputs B and C.

How many actions, operators are there in the math system? We start with two binary variables, A and B and make a truth table. A couple of binary functions are added to show how this works. This was first done in
Neurons for Robots as a basis for neural functions. Here we use it in a purely mathematical sense.

In	Out1	Out2
A B	AND	OR
0 0	0	0
0 1	0	1
1 0	0	1
1 1	1	1

We have Out1 = AND(A,B) and Out2 = OR(A,B) and note that the pair Out1 and Out2 is also a number.
AND is 1 (0001) and OR is 7 (0111).

Name the A,B pairs.

A	B	
0	0	Z
0	1	L
1	0	G
1	1	E

Make a number from the pair names. Make
names for the numbers.

Z	L	G	E	Logic	Arithmetic	comment
0	0	0	0	Z	Not(1)	ZERO
0	0	0	1	AND	*,/,MIN	E
0	0	1	0	GT	>	G
0	0	1	1	GTE	>=	GE
0	1	0	0	LT	<	L
0	1	0	1	LTE	<=	LE
0	1	1	0	XOR	<>,-	LG
0	1	1	1	OR	+,MAX	LGE
1	0	0	0	NOR	Not OR	Z
1	0	0	1	NXOR	Not XOR	ZE
1	0	1	0	NLTE	Not LTE	ZG
1	0	1	1	NLT	Not LT	ZGE
1	1	0	0	NGTE	Not GTE	ZL
1	1	0	1	NGT	Not GT	ZLE
1	1	1	0	NAND	Not AND	ZLG
1	1	1	1	E	Not LGE	ZLGE

If this is continued, we name each 4 bit number, the size of the next set of numbers we have to name is 65535. All of them, except Z, L, G and E are redundant.

The 0 action is NOT(1)
0000 NOT(1111) ; all ops produce 0

The primitive actions are one bit. They
are:
1 0001 AND, =, A = B, A * B,A / B, MIN(A,B)
2 0010   GT, >, A > B
4 0100 LT, <, A < B
8 1000 NOR, NOT(OR)

The secondary actions are 2 bits:
3   0011 GTE, >=, A GTE B,   GT + AND
5   0101 LTE, <=, A LTE B,   LT + AND
6   0110 XOR, <>, A XOR B,   GT + LT
9   1001 IS, A = B, NOT(XOR), NOR + AND

3 bit actions are:
7   0111 OR, +, MAX(A,B)
11   1011 ISGE,   IS + GT
13   1101 ISLTE,   IS + LT
14   1110 NAND, NOT(AND) IS + LTE

The 4 bit action is:
15   1111 ISOR, NOT(0000) ;All ops produce
1.

Now the actions can be identified in terms
of ZLGE in numerical order. This is how
they are identified, not how they act.

```
                               ;    ZLGE
Z                              ;0   0000
AND = E*N(Z)*N(L)*N(G)         ;1   0001
GT = G*N(Z)*N(L)*N(E)          ;2   0010
GTE = G*E*NOT(Z*L*E)           ;3   0011
LT = L*NOT(Z*G*E)              ;4   0100
LTE = L*E*NOT(Z*G)             ;5   0101
XOR = L*G*NOT(Z*E)             ;6   0110
OR = L*G*E*NOT(Z)              ;7   0111
NOR = Z*NOT(L*G*E)             ;8   1000
NXOR = Z*E*NOT(L*G)            ;9   1001
NLTE = Z*G*NOT(L*E)            ;10  1010
NLT = Z*G*E*NOT(L)             ;11  1011
NGTE = Z*L*NOT(G*E)            ;12  1100
NGT = Z*L*E*NOT(G)             ;13  1101
NAND = Z*L*G*NOT(E)            ;14  1110
NZ = Z*L*G*E                   ;15  1111
```

This set makes an algebra. NLTE – NXOR
= AND for example. NZ – AND = NAND.
We could make a machine for all 16
operators or we can make a machine with the
primitive operators and combine them to
make the rest of the 16 operators. This is
logic only here, so we go on to create
arithmetic.
Logic = {Not, LT, GT, EQ, NOR} = {0 2 4 8}

Not is a single input machine that is Out = Not(In). The machine is:

```
Not
 M = 1            ;M is last not 0 In
Loop
 Read In          ;Port In to variable In
 If In = 0, GOTO Yes, GOTO No
yes               ;In = 0
 Out = M          ;Not In
 GOTO Done
No                ;In <> 0
 M = In           ;set M
 Out = 0          ;Not In
Done
 Write Out        ;Out to Port Out
 GOTO Loop
```

In	Out
0	1
1	0

This Not node remembers the last not
condition. It is still
Out = Not(In) but it is not binary.
Not(0) = M
Not(M) = 0 and M = In
Now when we write X = Not(Y) we are using
this program.

LT is A less than B in the primitive operators. This has not been used as a function, as far as I know. To make a function of it, it must be a conditional function. That is, it must detect the condition before doing the function. All conditional nodes can have memory of the condition. Let the not condition of the function be 0. The memory not condition is M, where M is set during the last condition.

Here is the node.

```
LT
 M = 0
Loop
 Read L
 Read R
 IF L < R, GOTO Yes, GOTO No
Yes              ;L < R
 Out = L - R    ;tell us how much less
 M = Out
 GOTO Done
No
 Out = 0        ;M if you are was lt node
Done
 Write Out
 GOTO Loop
```

L	R	Out
0	0	0
0	1	-1
1	0	1
1	1	0

```
Z = X LT Y or Z = LT(X,Y)
Z = X WLT Y or Z = WLT(X,Y)
```

Greater than, GT or EQ differs only in the condition.

```
GT
 M = 0
Loop
 Read L
 Read R
 IF L > R, GOTO Yes, GOTO No
Yes
 Out = L - R    ;tell us how much
 M = Out
 GOTO Done
No
 Out = 0        ;M if wasgt
Done
 Write Out
 GOTO Loop
```

```
L R   Out
0 0   0
0 1   0
1 0   1
1 1   0

Z = X GT Y or Z = GT(X,Y)
Z = X WGT Y or Z = WGT(X,Y)

Nor
Loop
 Read L
 Read R
 Out = COM(L OR R)  ;Nor = Not(OR)
 Write Out
 GOTO Loop


L R OR Out
0 0 0  1
0 1 1  0
1 0 1  0
1 1 1  0
```

Now we have Not, LT, GT and NOR as nodes and the pattern for WLT and WGT condition memory nodes to come.

We have EQ left to do. The condition for EQ is =. That is, L = R. This is so much like IS that I use it for the node name. L IS R. Since the condition is true or false, we must consider the case when both L and R are 0. Since they are equal, this is a true condition, so we must provide a true answer, since we cannot use L. So there is an additional condition to test for here.

```
IS
 M = 0
Loop
 Read L
 Read R
 IF L = R, GOTO Yes, GOTO No
Yes
 Out = L          ;send out magnitude
 M = L            ;set memory
 IF Out = 0, GOTO Zero, GOTO Done
Zero
 Out = 1
 GOTO Done
No
 Out = 0          ;or M in WAS node
Done
 Write Out
 GOTO Loop
END IS

Z = X is Y or Z = IS(X,Y)
Z = X WAS Y or Z = WAS(X,Y)
```

And we have the pattern for the WAS node to come.

There is no easy binary way to establish the arithmetic operators, so we start with known operators and let the others come to us as needed. All of these operations are continuous, not logical, functions. Continuous functions have no memory.

Already, we have Plus (+), Minus (-), MPY (*), DIV (/). Arith = {+, -, *, /, MAX, MIN}

The pattern for continuous functions is:

```
Arith
Loop
 Read L
 Read R
 Out = f(L,R)  ;+, -, *, /, MAX, MIN
 Write Out
 GOTO Loop
END Arith
```

```
;The P node (+)
P
Loop
 Read L
 Read R
 Out = L + R
 Write Out
 GOTO Loop
END P

Out = L + R,  Z = X + Y

;The C Node (-)
C
Loop
 Read L
 Read R
 Out = L - R
 Write Out
 GOTO Loop
End C

Out = L - R,  Z = X - Y
```

MAX is alright for the function name. For
the infix I want smaller names, so I chose
o for MAX and a for MIN

Here are the nodes:

```
o
Loop
 Read L
 Read R
 IF L >= R, GOTO Yes, GOTO No ;(MAX)
Yes
 Out = L
 GOTO Done
No
 Out = R
Done
 Write Out
 GOTO Loop
END o
```

Out = I o R, Z = X o Y

```
a
Loop
 Read L
 Read R
 IF L <= R, GOTO Yes, GOTO No ;  (MIN)
Yes
 Out = L
 GOTO Done
No
 Out = R
Done
 Write Out
 GOTO Loop
END a

Out = L a R, Z = X a Y
```

Multiply and divide are shown as multiple additions and subtractions. First, multiply is:

```
MPY
 M = 0
Loop
 Read L
 Read R
 IF L < R, GOTO Yes, GOTO No
Yes
 M = M + R      ;add R L times
 L = L - 1
 IF L <= 0, GOTO Done, GOTO Yes
No
 M = M + L      ;add L R times
 R = R - 1
 IF R <= 0, GOTO Done, GOTO No
Done
 Out = M
 M = 0
 Write Out
 GOTO Loop
END MPY

Z = X * Y
```

```
DIV
 M = 0
Loop
 M = 0
 Read L
 Read R
 M = L
 IF L = 0, GOTO Zero, GOTO No
No
 M = M - R
 Out = Out + 1
 IF M <= 0, GOTO Yes, GOTO No
Yes
 GOTO Done
Zero
 Out = 0
Done
 Write Out
 GOTO Loop
END DIV
```

Out = L / R ; or Out = L DIV R

With proper scaling, DIV can compute trig
functions.

Both MPY and DIV count. MPY counts additions and DIV counts the number of subtractions. MPY counts by subtracting the count and DIV counts by adding the count. Note that we can divide by 0 here by making the result 0. All the functions are developed in Hal Algebra following.

Book 4
Hal Algebra

Introduction

Mathematics does not provide all the tools we need to construct or describe complex machines like robots, animachines and living beings. We are now moving from digital to numeric tools since binary digital tools are not adequate for anything but simple machines like computers.

Mathematical systems are languages that must be learned. To appreciate them you have to play with them. This is a simple system that will let you play with making computers into math machines. This system uses old mathematical words like: number, base, sets, variables, systems, functions, counting, arithmetic and algebra. It introduces new mathematical words like: uncounting, objects, actions, sensors, motors, nodes, cables, plugs, sockets, data streams and HalTrees.

The assumption underlying all this is: any data can be reduced to number or coded by a number. (There are more numbers than objects, so that is not a restriction. We can credit Pythagoras with this notion, although he would have applied it to object construction also.) The restriction applied is: any function or algebraic nodes must be reproducible in working hardware. If it cannot be built it will not be a function in this system. All that means in practice is that all nodes produce and work on data streams only.

Inputs and Outputs

Inputs are relative data coming in via an input port. Domain is the set of inputs. An input port allows us to read data from outside. Read X implies reading from port X to the variable X.

Outputs are relative data going out via an output port. Range is the output. Write X implies moving the variable X to an output port. An output port allows us to send data out to another node or whatever.

Inputs and outputs are relative because they depend on viewpoint. Your output might be the next lower node's input. Your input might be the last higher node's output.

Procedural and calculated Definitions

```
Let A = B - C
     A procedural definition is the
defining of the = and minus machines. A is
the output of B - C. B and C are inputs
relative to this equation.
Procedural definition of A = B - C
Loop
 Read B
 Read C
 A = B - C
 Write A
 goto Loop
```

A calculated definition is when you
define one variable in terms of other
variables.
A = B - C defines A in terms of B and C.

B = E + F defines B in terms of E and F
C = G + H and these definitions grow by two
each time until you get to raw data input
that does not need to be defined.

Variables and Sets

Variables are containers of number.
Variables have a name and content. X = 2 is
a variable with a value.

Sets are containers of objects. Sets
have a name and content. X = {a, b, c, d}
is a set of 4 things.

Since both sets and variables have the
same attributes, they can be used
interchangeably. A variable is a set with
one object.

Commentary:

A variable is a box. A set is a bag of
objects that may be variables.
To create either variables or sets, you
give them a name. Sets or variables with
the same name are the same sets or
variables.
1. A = B + C and
2. B = D + E means B is connected in the
two equations. The output of 2. is an input
to 1. 1. can be an input to some other
equation.

Numbers and Number symbols

Numbers are not visible. Number symbols are all that we can see. 0 1 2 3 is a sequence of number symbols. The actual symbols used are arbitrary, but they should represent a progression. All numbers in this system are cyclic digits. Numbers contain a reverse ordered, connected set of digits. N = d[n] + d[n-1] ... + d[1] + d[0].

d[n] = dn*b^n where dn < b. (We do not have to think about this often when we use a calculator.)

Number base and Digits

The count of number symbols plus the zero symbol is the number base. Let that number be the variable b. b is the number of symbols required to write all the symbols of a digit. Thus b is the number of possible numbers in the digit. Write the symbol sequence 0 1 2 3. If that is the digit, then b is at least 4. Observe the last symbol and add one for the zero. Or just count the symbols, starting at zero.

Objects and Actions

An object is any non active thing. Every noun is an object. Objects can have properties. An object concept is an object plus all its known properties. Of course number symbols are objects, since, by themselves, they do nothing.

An action is any active process. Every verb is an action. Operators perform actions on objects. Actions are sometimes used as objects. As we have seen, there is an algebra of AND, OR, <, >, and so on.

Systems and Functions

A system is a coordinated set of actions and objects.
This System = {These objects, These actions} Outside the system definition is the program that relates actions to objects. The general relation is
system = Tree(actions[t](objects[2*t]))

Actions work on or with objects. A function is a mapping of the objects of one set (the domain or input) to those of another set (the target, the range or the output). The mapping is many to one, never one going to many. $O = f(L,R)$ is a function of L and R. A function is to an action as a variable is to a number. It is a place holder for an equation while a variable is a place holder for a number.

Computer and Program

A computer is a device that reads a program and executes it. This computer can read data and write data, so it can be a function. This computer can do any function of data read to data written. Included in the actions of the computer there is at least a GOTO code and memory registers for the input data and temporary or scratch memory and conditional (IF, Then) ability. Computer = {program, read, write, functions, goto, if-then}

A program is the sequence of actions that the computer function should do on the input data to perform the function that goes to the output datum. Included in the program is the ability to recognize labels from code.

```
Label
 Code1
 Code2
Label2
 Code3
 ...
```

Counting and Uncounting

With one object or number, about the only thing you can do is to count from it or uncount to it.

Counting is a function of one variable. c = count(k).
Counting is almost integration. The code for counting is:

```
Count
 Read start
 Read stop
Loop
 start = start + 1
 If start >= stop, stop, goto loop
 Write start
 Goto Count
```

Uncount is a system of count and the function uncount.
uc = uncount(c). Uncount is almost differentiation.

```
UnCount
 Read start
 Read stop
Loop
 start = start - 1
 If start <= stop, stop, goto Loop
 Write start
 GoTo UnCount
```

Arithmetic and Algebra

Arithmetic is a system of two variables. A + B, A − B.
You can do the arithmetic but cannot pass it on.

Algebra is a system of three variables. C = A + B. Now you can store the result and pass it on. Algebra works on anything with a name and content. So far we have seen variables and sets with those attributes.

Sensors and Motors

Sensors read nature and write data streams. Sensors are new systems in mathematics. Anything that converts an aspect of nature and outputs a number is a sensor.

Motors read data streams and write nature. Anything that reads a data stream and converts it to an action is a motor.

A data stream is a stream of data such that if there is a datum at time t, there will be a new datum at time t+c forever. Any variable can become a data stream by attaching time. Given X, the data stream is X[t].

With reference to data streams, sensors are output devices and motors are input devices.

Nature -> Sensors -> Motors -> Nature
Sensors can connect directly to motors. Motors connect to sensors through nature or in a node.

Nodes and Cables

Nodes are actions or system. These are
the math machines we are building. A node
is a function with inputs (motors) and
output (sensor).

Cables carry the variables of node
functions. Cables carry data streams. Nodes
get data from cables and put data on cable.
The input and output ports on all nodes
connect with cables.

Cables keep the data streams private.
There are of course radio and light data
streams that need no cables. In that case
we make a radio in and radio out nodes to
read and write these data streams.

Plugs and Sockets

Cables have plugs. Nodes have sockets. Plugs connect to sockets. Any plug can connect to any socket or any number of sockets. A cable can be local or global, depending on its connections. A cable is a single variable.

A socket and plug together can take on any variable name.

Inside Hal Functions

We have examined the possible logic functions, the arithmetic and conditional functions. Here is a list of all the current Hal Nodes. This list is also in Neurons for Robots, in a slightly different form.

Creating HalStreams - Sensors

Nature -> **Sensors** -> Brain -> Motors -> Nature

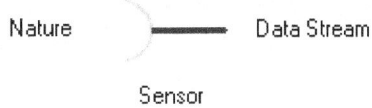

Nature   Data Stream

Sensor

Figure 3   Out = f(Nature), Sensors

The job of a sensor is to create and write a Hal stream from nature. Sensors decode and abstract from nature. A sensor sits half in nature and half in a brain, at the limits of the body. The sensor senses one item from nature. It is the fetch machine for this Hal stream. Your output is the axon if you are a neuron.

A sensor has a window to some feature of nature. It has a calculator. (All Hal stream manipulators have calculators and data storage registers if needed. Hal stream creators and consumers have all the tools of technical history available as required.)

The operation of a sensor is:

113

1. Observe the sensor quantity through the window.
2. Compute the number **O(t)**. Make it an integer.
3. Write the number to Out
4. Go to 1.   ; Cycle time is Cs.

Operation 1 observes nature. The means of observation can be electromagnetic, pressure, heat or inertial (light, sound, temperature or acceleration). Every HalNode is a virtual sensor since it creates a Hal stream.

Operation 2 can be complex. This is generally the process of converting an analog quantity into a number, as fast as possible. Sensors can be classified as to type and can be made universal with integer output. Then you can buy them off the shelf. The numbers produced by us as a sensor may be integers but you should not do anything else to them. No integration or differentiation should be done. Sensors have no memory. Sensors sit half in the world and half in the brain. In various applications, sensors encode and abstract from nature to some data type.

It takes a time, c, to create the
sensor Hal stream. Data can arrive in
nanoseconds, microseconds, milliseconds,
seconds, minutes, hours, days, months and
years. To make any Hal stream
differentiable you follow the rule of
keeping a datum available until there is a
new datum, regardless of the interval
between them. This makes the Hal stream
solid, with no intervals of bad data. If
the data is faster than you can pick it up
then you will skip some data and the output
Hal stream will be a sample of the incoming
data. If the data is very slow, then there
might be intervals of constant data.

The sensor adds time to the sensed
variable. The output Hal stream is two
dimensional, time is added to the data
stream and time is always implied in every
Hal stream variable.

The HalStreams emitted by a sensor can
be of any type.

k	constant
kN	Line of slope k
Nx	Combinatorial numbers of degree x
kPx	All polynomials of scale k and degree x
f(x,t)	Any function of a variable

In practice, you must be careful to
note the type of output you provide from
the sensor. In the analog world,

differentiation and integration can occur
by accident. Generally, a series capacitor
will differentiate. A parallel capacitor
will integrate. Your specifications should
specify if you do either.

Some sensor types are:
1. Temperature
2. Pressure
3. Light
4. Acceleration
5. Time
6. Numeric pad
7. Latitude and longitude
8. Radio
9. GPS

Units are those of length, mass, time, electric current, temperature and luminous intensity. Sensors should always be used in pairs. With pairs of sensors, you control another parameter.

A single sensor "sees" a single spot. A pair of sensors "sees" a single object. The distance between the sensors determines the minimum size of the object that can be detected. Increasing the distance increases the area covered and decreases discrimination. However, using another pair of sensors at a greater distance like this also increases the area covered while preserving size:

```
S1-S2---------------S3-S4 ; general
LL-LR---------------RL-RR ; left and right
```

There are nodes between sensors and other nodes, to convert units into numbers. The most general of these is the ADCNode, which converts voltage into number. There is also an RCNode which converts a radio control servo signal into an integer. Other types can be developed as needed.

The sensor part that looks at nature has a field of view. You measure field of view as a solid angle. A field of view can be represented by a cone radiating from the sensor. Everything inside the cone is on and everything outside is invisible or off. The angle of the cone can go from zero, completely closed to 180 degrees. (You can rotate to get the full 360 degrees, since it is symmetric.) The sensor has to be held by the body or body part. This by itself determines some of the field of view. Quite often you have to manipulate the field of view with a lens. Mainly you want to make it easy for the brain to see out there instead of in here.

What makes this important is that with multiple sensors, the fields overlap and form a pattern in space. This encodes a lot of information for the brain.

Mechanically, I have these sensor nodes:

1. NbrNode puts a constant on the data cable. At present it is 8 switches, which means it can be used for general switch detection.

2. ADCNode converts a voltage to a positive integer.

Working on HalStreams - HalNodes

Sensors feed data from outside to inside. All the nodes that follow, until we get to motors, belong to inside. There are two inside node platforms. TNodes have one input and one output. SNodes have two inputs and one output. TNodes make functions of a single Hal stream. SNodes make function trees and Hal Trees.

TNodes

We start with the single input, single output nodes. These are TNodes and are the hardware platform for TNodes, MoNodes, and SMNodes.

Keep this in mind for the descriptions and code that follows.

All TNodes Read I, Do function, Write O. Only functions differ for each node.

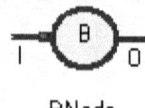

BNode

Figure 3 BNode

 We start the brain with the smallest
and simplest of HalNodes. In the brain
BNodes are both motor and sensor. They read
and write. They have synapse and axon. They
do not change dimension of the Hal stream.
Bnode's job is to catch the input and add
128 to the eight bit input number. This is
enough to convert -127 to 128 to 0 to 255
and backward, just like magic since we roll
the base 256 wheel over each time.

 Since we are working on a single Hal
stream it is easier to express this HalNode
as a function.
The HalNode equation is:
Out = b(I)
The BNode program is very simple:
BNode
Loop
 Read I
 Out = I + 128
 Write Out
 Go to Loop

This HalNode is used when your sensor is providing numbers from 0 to 255, all positive, and you must make an integer right away. Since a CNode always creates integers, this HalNode is seldom needed. If you have a dumb DAC then a BNode will assure it gets values from 0 to 256 as the DAC expects.

Negating HalStreams --- N Node

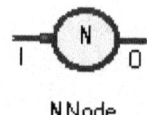

N Node

Figure 4 Out = N(I)

One more simple HalNode. NNode's job
is to take each input number and subtract
it from 0. Like the BNode above, the
negation works both ways. Again since we
are working on a single Hal stream we use a
functional notation. The NNode equation is:

Out = n(I) or **Out = -I** where there is no
possible confusion.

The NNode program is:
NNode
Loop
 Read I
 Out = 0 - I
 Write Out
 Go to Loop

A Function HalNode ---- Not Node

NotNode

Figure 5 Out = Not(I)

NotNode's job is to make binary logical reversals of meaning. It is a function that observes I. If I is zero, it sends out 1, else it sends out zero. A Not node is used to complete all the conditional HalNodes. Put it in front of a Z node and you have a not Z and so on. It is binary logical because it cares not for values except one and zero.

As assembler code:

```
NotNode
 OldI = 3
Loop
 Read I
 If I = 0 then OldI = I and Out = OldI else
Out = 0
 Write Out
 Go To Loop
```

As spreadsheet function. A2 is input.

```
=if(A2=0,A1,0)  ;cannot do OldI here.
```

This is not like the spread sheet NOT function, since the spread sheet returns True and False. Of course True is one and False is zero. This is not a Boolean NOT, it is an arithmetical Not.

```
NotNode truth table
 O = Not(I)
```

I	O
-1	0
-1	0
-1	0
0	-1
0	-1
0	-1
1	0
1	0
1	0

Note: M is used so we have a little bit of memory. This provides some redundancy for the Robot. If the input to NotNode is always 0, then NotNode defaults to a 3.

A Function HalNode ---- ABS

ABS Node

Figure 6   Out = ABS(I)

ABSNode is an absolute function. Sometimes we are not interested in direction, only in magnitude. The node determines the sign on the I input and makes Out positive.

```
ABS
 Read I
 If I < 0 then send complement of I to O
else O = I
 Write O
 GOTO ABS
```

Suppose you have rudder and propeller. You have sensor L and sensor R. rudder = L - R If we only want to move forward then let propeller = ABS(rudder)

L	R	rudder	propeller
0	0	0	0
0	1	-1	1
1	0	1	1
1	1	0	0

Differencing a Hal stream - -- D Node

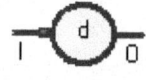

DNode

Figure 7   Out = d(I)

You are a DNode, a single input HalNode.

The job of a differencing function is to subtract each incoming datum from the Hal stream, and send the difference out as a new Hal stream. A curve comes into us and we emit the instructions for making that curve. $d(x) = dx/dt$ because $dt = 1$ in every Hal stream. You read I and produce d(I).

O = d(I) ;Example: Acceleration = d(Velocity)

```
DNode
Start
 OldIn = 0
Loop
 Read I
 Out = I - OldIn
 Write Out
 OldIn = I
 Go To Loop
```

To difference by hand:
Write down the Hal stream sequence like this:

```
j =      1 2 3 4…
DS =     2 4 6 8…
d(DS) =   2 2 2…
```

Note: $d(DS)[j] = DS[j] - DS[j-1]$ for every j

$O[j] = I[j] - I[j-1]$; $j > 0$

In a calculator:
1. Enter the jth number. (DS[j])
2. Subtract the j-1th number. (DS[j] - DS[j-1])
3. Record the subtraction.
4. Move to the next number. (j=j+1)
5. Go to 2.

In a spreadsheet you can only difference a column or row. Suppose you have a column of numbers in B. Then d(B) would be the spreadsheet column beside it:
= B2 - B1 ; replicated down its own column.

In truth tables HalStreams are written going down in a column. In the DNode, think of I as a Hal stream, not as a number.

DNode truth table

I	O
1	x	; I is a triangular wave here.
2	1
3	1
2	-1
1	-1

To difference, you must also have memory, but only for one past datum. Because the derivative of a constant is zero, a DNode strips off offsets. The derivative of … 25,26… is the same as the derivative of …1,2… To obtain a derivative you must have at least two numbers. The offset that you throw away is the very first number in the Hal stream. Real DNodes produce continuous differences of their input HalStreams and any idea of a first number is lost. Every Hal stream can be differenced as long as there are two numbers available.

In the domain of Hal streams, DNodes obey all the rules of calculus without the restriction of requiring a limiting process.

```
C1 = d(C0)
C2 = d(C1)
C3 = d(C2)
C4 = d(C3)
```

You can also express this as:
$k = d5(C0)$; saying k is the 5th derivative of C0
Note that $d5(C0) = C4 - 4*C3 + 6*C2 - 4*C1 + C0$ and that we need 5 numbers to do that. So $dn(x)$ requires n numbers.

A DNode is also a virtual motor and a
virtual sensor, so those properties attach
to it also.

Integrating a Hal stream ----- I Node

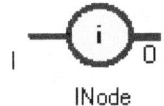

INode

Figure 8 Out = i(I)

The incoming Hal stream I is a set of instructions on how to make a curve. The instructions are: go up so much, stay here, or go down so much. The result of this is to build a curve, generally from another curve.

The job of an integrator is to add each incoming datum from the Hal stream to a count, and send the count out as a new Hal stream. Any integration needs a constant simply because the differential of a constant is zero and we obey the rules of calculus. In a Hal stream, additions and subtractions of constants produce a phase shift. To use a constant with an INode, use it outside like Out = i(I) + k

Out = i(I) ; Example : Velocity = i(Acceleration)

i(I)[j] = i(I)[j-1] + I[j]

The operations are:

1. Look at the last Hal stream integral **O[j-1]**.
2. Add the input Hal stream datum **I[j]**.
3. Write **O[j] = O[j-1] + I[j]**
4. **O[j-1] = O[j]**
5. Go to 1. ; Cycle time is Ci.

As assembler code:

```
BEGIN i
 Out = 0
Loop
 Read I
 Out = Out + I   ;the integration
 Write Out
 Go To Loop
END i
```

Integrating with a calculator: The calculator should have memory. If the calculator has no memory, store to memory would mean write to paper.

1. Set the starting constant in memory.
   (Enter the constant. Store it in memory.)
2. Enter the next number in the Hal stream.
3. Add memory.
4. Store to memory
5. Go to 2.

In a spreadsheet, you will have to set up a column or row to integrate. Suppose the column B is to be integrated. Set up the integrator column in C as:
=C1 + B1 ; replicate that down C. Note that C1 is the integration constant.

As an INode, you must have memory. An INode can count to a number larger than its bit capacity. Like all adders, you simply let it roll. If last O was 200 and I = 150, then new O would be 94 (350 -256).

INodes obey all the rules of calculus without the limitation of requiring a limiting process. kPD = i(kP(D-1)

X = d(i(X)) and N = i(d(N))

Every integrator increases the degree of the incoming Hal stream. The integrator uses the differential of the curve being built as instructions to build the curve.

INode truth tables

I	Out	I	Out
1	1	1	1
1	2	2	3
1	3	3	6

```
C1 = i(1)        ; C1 = 1 2 3 4 5 ...
C2 = i(C1)       ; C2 = 1 3 6 10 15 ...
C3 = i(C2)       ; C3 = 1 4 10 20 35 ...
...
```

And so on. This string of INodes computes every combinatorial sequence. C1 is combinations of N taken 1 at a time. C2 is combinations of N + 2 taken 2 at a time, and so on, to combinations of N + n + k taken k at a time. Note that all n things are produced and we are manipulating k only.

Since an integrator is a counter, INodes can be used whenever you need to time or count some event or condition. Let s = A z 1 ; s is one if A is zero, else A is zero.

Now timeofs = i(s) ; will count while A is zero and will retain that, since i(0) = x. If A is zero for 5 c times, then timeofs will be 1 2 3 4 5 5 5 5 5 5 5 5 5 ... This is a cumulative time, since timeofs does not reset. If you need frequency, become an FNode.

If you need an integration constant, add it externally. For example, i(I) + c = 0 ; is integral with constant.

Feed i(I) into a P node adds another Hal stream and can be any integration constant. Taking the constant out of the node gives you maximum flexibility.

An accumulating function - PLUS

Figure 9 PLUS node

PLUSNode is a node that accumulates.
Out = Out + I ; is the action. It is almost
an I node.

I	Out
0	0
1	1
2	3
2	3
1	4

The difference is you add only if the input
is different from the last input.

A decreasing function --- MINUS

Figure 10   MINUS node

MinusNode is a node that decreases itself. It is the inverse of the PLUS node. It is almost a D node. It is almost an un counter.

Out = Out - I ; is the action.

I	Out
1	-1
2	-3
2	-3
1	-4

This is the end of the single input
nodes. Now we start with two input nodes.

SNodes

These create Hal Trees.

All SNodes read L then read R, do a
function and write Out. Only the function
code is shown for the functions. There is
code that is common to all SNodes. The
routines are [Init], [Read L and R], and
[Write Out]. For reference all these are
shown here.

Arithmetic or stream Algebra nodes

Adding Hal Streams --- P node

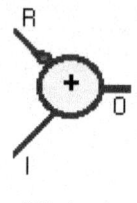

PNode

Figure 11 Out = L + R

Out = L + R. it does the + operation and the = operation. It has two motors and a sensor. You have two synapses and an axon.

Inside the P node, you have two numeric displays, one labeled **L** and the other labeled **R**. **L** and **R** come from sensors, real or virtual. To you, L and R are motors that read HalStreams. You, the P node, cannot know whether the HalStreams originate from sensors or other nodes. Your job is to mash two HalStreams together and produce the Hal stream Out. The operation of a P node is:

Out = L + R   ;Out[Future] = I[Present] + R[Present]

142

1. Observe the **L[Present]** Hal stream datum.
2. Add the **R[Present]** Hal stream datum.
3. Write **Out[Future] = L[Present] + R[Present]**.
1. Go to 1. Cycle time is Cp.

As assembler code:

```
BEGIN P
Loop
 Read L
 Read R
 Out = L + R
 Write Out
 Go To Loop
END P
```

In a spreadsheet you can add any cell to any other cell.
= A2 + G3

PNode truth tables

L	R	O
-1	-1	-2
-1	0	-1
-1	1	0
0	-1	-1
0	0	0
0	1	1
1	-1	0
1	0	1
1	1	2

L	R	O
0	0	0
0	1	1
0	2	2
1	0	1
1	1	2
1	2	3
2	0	2
2	1	3
2	2	4

A PNode, as with every two input HalNode, is two virtual motors, from reading two HalStreams, and a virtual sensor from writing a HalStream. Here, in the brain, all HalStreams can be assumed abstract and dimensionless. It should be noted that L and R are instructions from nature, which includes the rest of the brain.

As a P node, you can compute a number larger than our bit capacity. If that happens then you simply let the wheel roll and the datum on the O Hal stream will be O - b. Example:
L = 200 and R = 150. Then O = 94. (350 - 256)

Combine a P node with an NNode and you can make a C node.

Comparing HalStreams - C node

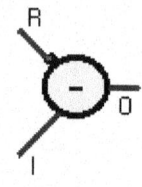

CNode

Figure 12 Out = L - R

You are a C node.

The job of a CNode is to compare two HalStreams. Like the P node, it has two numeric displays that show the HalStreams it is watching. The actions are:

Out = L - R   ;Out[Future] = L[Present] - R[Present]

1. Observe the Hal stream datum **L**.
2. Subtract the Hal stream datum **R**.
3. Write the Hal stream datum **Out = L - R**.
4. Go to 1.   ; Cycle time is Cc.

As assembler code:
CNode

```
Loop
 Read L
 Read R
 Out = L - R
 Write Out
 Go To Loop
```

In a spreadsheet you can subtract any cell
from any other cell.
= G3 - A2

CNode truth tables

L	R	O		L	R	O
0	0	0		-1	-1	0
0	1	-1		-1	0	-1
0	2	-2		-1	1	-2
1	0	1		0	-1	1
1	1	0		0	0	0
1	2	-1		0	1	-1
2	0	2		1	-1	2
2	1	1		1	0	1
2	2	0		1	1	0

C node gets the same result whether it is fed an integer or an offset binary pattern. So it converts everything into an integer.

As in all two input nodes, a C node is two virtual motors and a virtual sensor or two virtual synapses and a virtual axon.

All the theorems of Boolean algebra can be extended to work on HalStreams without restriction. (HalStreams replace bits and arithmetic replaces logic. Unfortunately most switching algebra is to compensate for bit limitations, so very little applies to HalStreams. However work has not even started on ruining Hal algebra.)

Subtraction, like differentiation can change Hal stream types. Subtraction, in a Hal stream is spatial differentiation. This gives us another view of differentiation.

We are accustomed to differentiation in time. In space, differentiation covers length. The first derivative looks at one spot, the second looks at two spots and the nth looks at n spots. When CNodes make HalTrees, each column of the tree is the differential of the column before it, since time is along the columns.

There is no exact mechanical analogy for any C node. Mechanics requires units. You can make mechanical torque differentials or force differentials and so on, but there are no abstract mechanisms that can combine into HalTrees.

To see how a C node works, automate a number wheel. Fix it so you can put in a number, and the wheel will rotate to match the number. Create two sensors, one to read the number on the wheel (W), the other to read the number you put in (K).

Make a motor to run the clicker on the wheel (Click). The equation is: Click = K - W

Suppose W is sitting on 5 and you put in 9. Click starts at 4 and starts clicking. You need 4 clicks to get from 5 to 9. But a CNode is more subtle than that.

The first click is from 5 to 6. Now Click is at 3. New instructions have been supplied. That is the effect of continuous action.

Click	K	W	Action
4	9	5	One click
3	9	6	One click
2	9	7	One click
1	9	8	One click
0	9	9	Done

The C node, when in a feedback situation will always count down like this. It will ignore ambient changes in inputs and its last instruction to a motor is always 1 and then 0 to stop.

Choosing the Largest Hal stream -- O Node

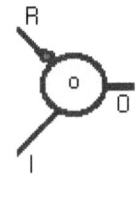

ONode

Figure 13   Out = L o R

The job of an ONode is to compare two Hal streams and create a new Hal stream with the largest data from either. This is a fuzzy logic OR operation applied to HalStreams instead of sets.

Out = L o R ; Out is equal to L or R

1. Observe the L Hal stream datum
2. Observe the R Hal stream datum
3. Choose the largest
4. Write it to the Out Hal stream
5. Go to 1.

As assembler code:

ONode
Loop

```
Read L
Read R
If L > R then Out = L else Out = R
Write Out
Go To Loop
```

In a spreadsheet you can determine the max
of any set of cells.
```
=MAX(A2,C2,G3)
```
or
```
=MAX(A2:A42)
```

ONode truth table b = 3

L	R	Out
0	0	0
0	1	1
0	2	2
1	0	1
1	1	1
1	2	2
2	0	2
2	1	2
2	2	2

ONodes are a convenience. Sometimes you need to know a value or to lock in a maximum value. If you need to know the single largest of a number of sensors, use a sparse tree of ONodes. A HalTree of ONodes will pick the largest number of any number of inputs, always and continually.

Choosing the Smallest Hal stream - A Node

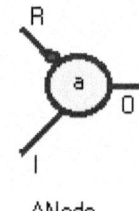

R

a   0

I

ANode

Figure 14   Out = L a R

The job of an ANode is to compare two Hal streams and create a new Hal stream with the smallest data from either. This is a fuzzy logic AND operation, again using HalStreams instead of sets.

Out = L a R ; Out is equal to L and R

1. Observe the L Hal stream datum
2. Observe the R Hal stream datum
3. Choose the smallest
4. Write it to the Out Hal stream
5. Go to 1.

As assembler code:
ANode
Loop
 Read L

```
Read R
If L < R then Out = L else Out = R
Write Out
Go To Loop
```

As a spreadsheet function, use the MIN
function.
```
=MIN(A2:A43)
```

ANode truth table (b = 3)

L	R	O
0	0	0
0	1	0
0	2	0
1	0	0
1	1	1
1	2	1
2	0	0
2	1	1
2	2	2

If you need to know the smallest value of a large number of sensors, use a sparse tree of ANodes. It is interesting to note that ANodes will do chemical arithmetic. Given quantities of 2H and O then quantities of water = 2H a O. The quantity of water is measured and limited by the minimum quantities of 2H or O.

Multiplying Hal Streams - MPY node

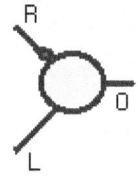

MPY node

Figure 15   Out = L * R

MPYNode is a node that multiplies R
and L.
Out = L * R ;
MPY is included for completion.

L	R	Out
0	0	0
0	1	0
1	0	0
1	1	1

MPY is the same as AND on binary numbers.

Dividing Hal Streams ---- DIV node

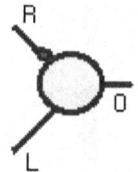

DIV node

Figure 16   Out = L / R

DIVNode is a node that divides L by R
Out = L / R

DIV does an integer divide. It is included
for completion of the arithmetic.

L	R	Out
0	0	0
0	1	0
1	0	0
1	1	1

On binary numbers DIV cannot be
distinguished from MPY. Note that L/0 = 0
here. (The function would not end if we did
not detect divide by 0. The way the code is
written, DIV would = 1 but there would
never be an output, so you would never see
it.)

A Boolean Logic Hal Node - AND Node

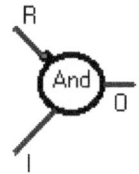

AndNode

Figure 17   Out = L and R

AndNode is a Boolean logic node. You read L and do an and
operation with the mask in R.

Operation
Out = L and R

Loop
 Read L
 Read R
 do L and R
 Write Out
 Go to Loop

Truth table

L	R	Out
0	0	0

```
0    1    0
1    0    0
1    1    1
```

It is convenient to code several
things into one number in the detection
process. Then use the AndNode to decode.

A Boolean Logic Hal Node - OR Node

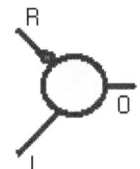

Figure 18   Out = L or R

When you want to add without carry, use OR instead of Plus.

```
Operation
Out = L or R

Loop
 Read L
 Read R
 do L or R
 Write Out
 Go to Loop
```

Truth table

L	R	Out
0	0	0
0	1	1
1	0	1
1	1	1

Here is the OR function code:

A Boolean Logic Hal Node - NOR

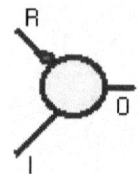

Figure 19   Out = L nor R

NorNode is a Not Or node. You are a unique logic function that detects equal and zero. NOR is the complement of OR.
Operation:
Out = L nor R

```
Loop
 Read L
 Read R
 Out = L or R
 Out = complement of Out ;0 -> 1 -> 0
 Write Out
 Go to Loop
```

Truth table

L	R	Out
0	0	1
0	1	0
1	0	0
1	1	0

A Boolean Logic Hal Node - XOR

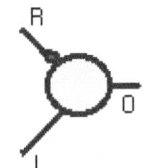

Figure 20   Out = L XOR R

    XOR is a unique logic function that detects less than and greater than.

Operation
Out = L xor R

Loop
 Read L
 Read R
 Out = L xor R
 Write Out
 Go to Loop

Truth table

L	R	Out
0	0	0
0	1	1
1	0	1
1	1	0

XOR is subtraction without carry.

Conditional algebra nodes

A Detection HalNode - IS Node

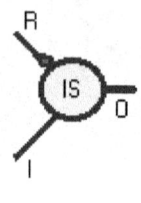

ISNode

Figure 21   Out = L is R

ISNodes observe the L and R inputs and if L is equal to R, then make Out = L, else make Out = 0.

Out = L is R ; Out equals L while L equals R, else Out = 0
"is" was chosen as the operator symbol because I do not want two = signs in an equation and it is language equivalent. To say x is y is the same as saying x = y. Of course is means
L = R, so we really should say iseq, since we will have isgt and islt for the conditions L > R and L < R respectively.

1. Observe the L Hal stream

2. Compare with the R Hal stream
3. If L = R then make Out = L Else make Out
= 0
4. Write Out
5. Go to 1.

As assembler code:
```
BEGIN IS
Loop
 Read L
 Read R
 If L = R then Out = L else Out = 0
 Write Out
 Go To Loop
END IS
```

Spread sheet form

IF(L=R, R, 0) If L is in A2 and R is in B2
then the spreadsheet command is:
=IF(A2=B2,A2,0)

IS Node truth table

L	R	Out
0	0	0
0	1	0
0	2	0
1	0	0
1	1	1
1	2	0
2	0	0
2	1	0
2	2	2

A logic HalNode is used for Hal stream
logic. They are used when you want an event
to occur when some threshold is passed.
Hal algebra is an algebra, so most the
rules of algebra can be used.
You should note that the ISNode is a
combination CNode and ZNode and should be
used when that combination is needed.

A Detection HalNode - GT Node

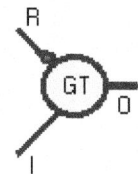

GTNode

Figure 22   Out = L gt R

GTNode observes the L input and if L greater than R, then make Out = L-R, else make Out = 0. GT is not a function in regular math.

Out = L gt R ; Out = L-R if L greater than R, else Out = 0
Out = L > R   ; alternate notation

1. Observe the L Hal stream
2. Compare with the R Hal stream
3. If L > R then make Out = L - R Else make Out = 0
4. Write O
5. Go to 1.

As assembler code:
BEGIN GT

```
Loop
  Read L
  Read R
  If L > R then Out = L - R else Out = 0
  Write Out
  Go To Loop
END GT
```

Spread sheet form
IF(L > R, L-R, 0) If L is in A2 and R is in B2 then the spreadsheet command is:
=IF(A2>B2,A2-B2,0)

GTNode truth table

L	R	Out
0	0	0
0	1	0
0	2	0
1	0	1
1	1	0
1	2	0
2	0	2
2	1	1
2	2	0

A Detection HalNode - LT Node

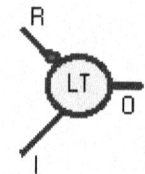

LTNode

Figure 23   Out = L lt R

LT node observes the L input and if L is less than R, then make Out = L-R, else make Out = 0. None of these conditional terms are functions in regular math.

Out = L lt R ; Out equals L-R if L less than R, else Out = 0
Out = L < R ; is alternate notation

1. Observe the L Hal stream
2. Compare with the R Hal stream
3. If L < R then make Out = L-R Else make Out = 0
4. Write Out
5. Go to 1.

As assembler code:

```
BEGIN LT
Loop
 Read L
 Read R
 If L < R then Out = L-R else Out = 0
 Write Out
 Go To Loop
END LT
```

Spread sheet form
IF(L < R, L-R, 0) If L is in A2 and R is in
B2 then the spreadsheet command is:
=IF(A2<B2,A2-B2,0)

LTNode truth table

L	R	Out
0	0	0
0	1	0
0	2	0
1	0	0
1	1	0
1	2	1
2	0	0
2	1	0
2	2	0

A Detection and memory Hal Node - WEQ

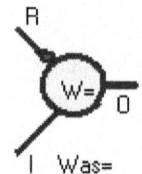

Figure 24   Out = L weq R

WEQ observes the L input and if L was R, then make Out = L and M = L, else make Out = M.

Out = L weq R; Out = L if L was R and M = L, else Out = M
WEQ certainly is not in regular math.

1. Observe the L Hal stream
2. Compare with the R Hal stream
3. If L = R then make M = L and make Out = M
4. Write Out
5. Go to 1.

As assembler code:
BEGIN WAS
 M = 0
Loop
 Read L
 Read R

```
 If L = R then M = L
 Out = M
 Write Out
 Go To Loop
END WAS
```

Spread sheet form
IF(L = R, L, M) If L is in A2 and R is in
B2 then the spreadsheet command is:
=IF(A2=B2,A2,A1)

WasEQ truth table

L	R	M	Out
0	0	0	0
0	1	0	0
0	2	0	0
1	0	0	0
1	1	1	1
1	2	1	1
2	0	1	1
2	1	1	1
2	2	1	1

A Detection and Memory Hal Node - WGT

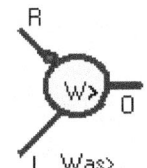

Figure 25   Out = L wgt R

Was greater than observes the L input and
if L was > R, then make Out = L-R and M = L-R,
 else make Out = M.
Out = L wasgt R ; Out equals L-R if L was > R
 and M = L-R and Out = M

1. Observe the L Hal stream
2. Compare with the R Hal stream
3. If L > R then make M = L-R and make Out = M
4. Write Out
5. Go to 1.

As assembler code:
BEGIN GT
 M = 0
Loop
 Read L

```
 Read R
 If L > R then M = L-R
 Out = M
 Write Out
 Go To Loop
END GT
```

Spread sheet form
IF(L > R, L, M) If L is in A2 and R is in
B2 then the spreadsheet command is:
=IF(A2>B2,A2-B2,A1)

WasGT truth table

L	R	M	Out
0	0	0	0
0	1	0	0
0	2	0	0
1	0	1	1
1	1	1	1
1	2	1	1
2	0	2	2
2	1	1	1
2	2	1	1

A Detection and Memory Hal Node - WLT

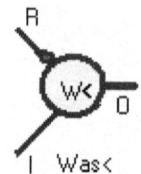

Figure 26   Out = L wlt R

Was less than observes the L input and
if L was < R, then make Out = L-R and M =
L-R,
 else make Out = M.
Out = L wlt R ; Out equals L-R if L was < R
 and M = L-R and O = M

1. Observe the L Hal stream
2. Compare with the R Hal stream
3. If L < R then make M = L-R and make Out
= M
4. Write Out
5. Go to 1.

As assembler code:
BEGIN WLT
 M = 0
Loop
 Read L
 Read R
 If L < R then M = L-R

```
 Out = M
 Write Out
 Go To Loop
END WLT
```

Spread sheet form
IF(L < R, L, M) If L is in A2 and R is in
B2 then the spreadsheet command is:
=IF(A2<B2,A2-B2,A1)

WasLT truth table

L	R	M	Out
0	0	0	0
0	1	-1	-1
0	2	-2	-2
1	0	-2	-2
1	1	-2	-2
1	2	-1	-1
2	0	-1	-1
2	1	-1	-1
2	2	-1	-1

Conditional Switching algebra nodes

A Conditional Switch Hal Node - L Node

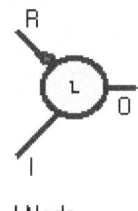

LNode

Figure 27   Out = L l R

LNode's job is to watch the L input and divert R into Out while it is less than zero.
Keep Out = 0 otherwise.

Out = L l R ; is the infix notation.

```
Start:
     Out = 0    ; Default status
     While L < 0 Out = R
     Go to Start
```

Assembler code is more like:
```
BEGIN L
Loop
 Read L
 Read R
```

```
 If L < 0 then Out = R else Out = 0
 Write Out
 Go To Loop
END L
```

Spread sheet form

IF(L<0, R, 0) Let L be in A2 and R be in B2. Then the spreadsheet command is:
=IF(A2<0,B2,0)

LNode truth table

```
Out = L 1 R
```

L	R	Out
-1	-1	-1
-1	0	0
-1	1	1
0	-1	0
0	0	0
0	1	0
1	-1	0
1	0	0
1	1	0

Note that R is unspecified. R can be anything from, a constant representing a code, to a complex Hal stream.

LNodes do not make combinatorial trees. An LNode is applied when a CNode is used as an object detector. Here a negative is the signal of detection. The LNode converts the signal of detection into a code of action. The first LNode says, "Here is the property of an object," if an object is detected, and says, "There is no object," otherwise.

All these nodes, Z, G and L are
conditional switches, switching data on
condition. They do not perform tree
functions.

A Conditional Switch Hal Node - G Node

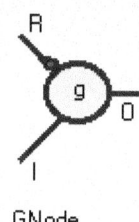

GNode

Figure 28   Out = L g R

GNode's job is to watch your L input
and divert R into Out while it is Greater
than zero.
Keep Out = 0 otherwise.
You are arithmetic logic, not Boolean
logic.

Out = L g R ; is the infix notation.

```
Start:
    Out = 0    ; Default status
    While L > 0 Out = R
    Go to Start
```

Assembler code is more like:
GNode
Loop
 Read L
 Read R
 If L > 0 then Out = R else Out = 0
 Write Out
 Go To Loop

Spread sheet form

IF(L>0, R, 0) Let L be in A2 and R be in
B2. Then the spreadsheet command is:
=IF(A2>0,B2,0)

GNode truth table

Out = L g R

L	R	Out
-1	-1	0
-1	0	0
-1	1	0
0	-1	0
0	0	0
END		
0	1	0
1	-1	-1
1	0	0
1	1	1

Note that R is unspecified. R can be anything from, a constant representing a code, to a complex Hal stream.

G Nodes do not make combinatorial trees. A G Node is applied when a CNode is used as an object detector. Here a positive is the signal of detection. The GNode converts the signal of detection into a code of action. The first GNode says, "Here is the property of an object," if an object is detected, and says, "There is no object," otherwise.

A Conditional Switch Hal Node - Z Node

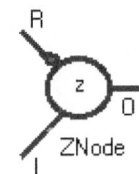

Figure 29   Out = L z R

ZNode's job is to watch the L input and divert R into Out while it is equal to zero.
Keep Out = 0 otherwise.

Out = L z R ; is the infix notation.

```
Start:
     Out = 0    ; Default status
     While L = 0 Out = R
     Go to Start
```

Assembler code is more like:
```
BEGIN Z
Loop
 Read L
 Read R
 If L = 0 then Out = R else Out = 0
 Write Out
 Go To Loop
```

END Z

Spread sheet form

IF(L=0, R, 0) Let L be in A2 and R be in
B2. Then the spreadsheet command is:
=IF(A2=0,B2,0)

Z Node truth table

Out = L z R

L	R	Out
-1	-1	0
-1	0	0
-1	1	0
0	-1	-1
0	0	0
0	1	1
1	-1	0
1	0	0
1	1	0

Note that R is unspecified. R can be anything from, a constant representing a code, to a complex Hal stream.

ZNodes do not make combinatorial trees. A ZNode is applied when a CNode is used as an object detector. Here a zero is the signal of detection. The Z Node converts the signal of detection into a code of action. The first Z Node says, "Here is the property of an object," if an object is detected, and says, "There is no object," otherwise.

A Conditional Switch Hal Node - DL Node

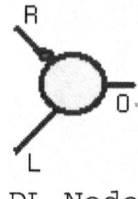

DL Node

Figure 30   Out = L dl R

A DL Nodes job is to watch the L input and divert R into Out while d(L) is less than zero.
Keep Out = 0 otherwise.
You are arithmetic logic, not Boolean logic.

Out = L dl R ; is the infix notation.

```
Start:
     Out = 0    ; Default status
     While d(L) < 0 Out = R
     Go to Start
```

Assembler code is more like:
```
DL Node
 OldL = 0
Loop
 Read L
```

```
Read R
DI = L - OldL
If DI < 0 then Out = R else Out = 0
Write Out
OldL = L
Go To Loop
```

DL Node truth table

Out = L dl R

L	R	DL	Out
-1	1	0	0
-1	1	0	0
-1	1	0	0
0	1	1	0
0	1	-1	1

A Conditional Switch Hal Node - DG Node

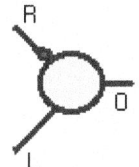

DG Node

Figure 31   Out = L dg R

A DG Node's job is to watch the L input and divert R into Out while d(L) is Greater than zero.
Keep Out = 0 otherwise.
You are arithmetic logic, not Boolean logic.

Out = L dg R ; is the infix notation.

```
Start:
     Out = 0    ; Default status
     While L > 0 Out = R
     Go to Start
```

Assembler code is more like:
```
GNode
OldL = 0
Loop
 Read L
```

```
Read R
DL = L - OldL
If DL > 0 then Out = R else Out = 0
Write Out
Go To Loop
```

DG Node truth table

Out = L dg R

L	R	DL	Out
-1	1	0	0
-1	1	0	0
-1	1	0	0
0	1	1	1
0	1	0	0

Note that R is unspecified. R can be anything from, a constant representing a code, to a complex Hal stream.

A Conditional Switch Hal Node - DZ Node

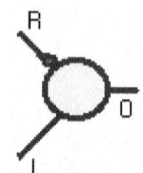

Figure 32 Out = L dz R

Out = 1 Done R

     A DZ Node's job is to watch the L input and divert R into Out while d(L) is equal to zero.
Keep Out = 0 otherwise.
You are arithmetic logic, not Boolean logic. This node can be used to compute when something on L is done. I had a node I called Done and it is the same as a dz node. So this is also a Done node.

Out = L g R ; is the infix notation.

```
Start:
    Out = 0    ; Default status
    While d(L) = 0 Out = R
    Go to Start
```

Assembler code is more like:
DZ Node
 M = 0

```
Loop
 Read L
 Read R
 If L - M = 0 then Out = R else Out = 0
 Write Out
 M = L
 Go To Loop
```

DZ Node truth table

```
Out = L dz R
```

L	R	d(L)	Out
-1	1	0	0
-1	1	0	1
-1	1	0	1
0	1	1	0
0	1	0	1

Note that R is unspecified. R can be anything from, a constant representing a code, to a complex Hal stream.

This is a reasonable set of nodes, but it can never be a complete set, since any computer program can become a node. So you build what you need. On the other hand, the CNode and NotNode can do most everything, just like AND and NOT or OR and NOT can make almost any Boolean algebra set. But you don't have to do that since you have a great selection of nodes to choose from.

Motors

Reading a Hal stream - Motors

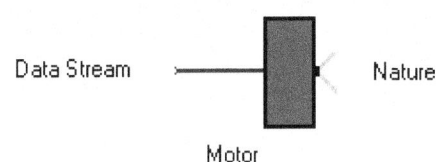
Data Stream   Nature

Motor

Figure 33   Motor = I

A real motor controller watches the I input and does what it says. It reads the data, putting units and meaning to the Hal stream. It is the decoder of this Hal stream. I is a synapse on a muscle, coming from some HalTree of some axon.

The job of a motor is to convert a Hal stream into an action. It is the final reader of a Hal stream although it does not consume it. The Hal stream can continue beyond the motor. But if a Hal stream ends, it ends with a motor. A HalTree ending in a motor is called a command tree. The size of the HalTree determines the possible actions of the motor. All that is connected to the motor is a function.

M = I[Present]

The operation of the motor is:
1. Observe the Hal stream.
2. Use the quantity to run an action in the direction and amount indicated by the number in the Hal stream.
3. Go to 1. ; Cycle time is Cm.

Assembler code is like:
Loop
 Read I
 DO I   ;with 8 bits you have 256 possible actions.
 Go To Loop

     Operation 2, or DO I, can be simple or complex and can only be specified when the design process is complete. Operation 2 must be specified. In the virtual motors of the other nodes, operation 2 is to simply put the input into a memory. In a CNode, for example, the R input motor makes a twos complement number of the number coming in on the cable. The I input simply puts the input number into a register.
     Motors require time to cycle. The speed of the action that is produced is usually slower than the cycle time of the Hal stream. Fortunately, nature has cycles of every scale and most actions of animals and Robots are very slow.
     The task of every motor is to decide what to do with the input integer. We can

make conventions here by always making minus numbers go left or counter clockwise or backward and so on. It is no big deal. You can always add an inverter to fix this.

It is in the motor that meaning is given to Hal stream variables. The purpose of the motor is to give expression to the semantic of the Hal stream variable. The construction of a motor that is a body part is still an art and requires all the technology available to the engineer. In time, you will be able to classify motors as to function so you can buy them off the shelf.

A real motor sits half in the brain and half in the world. A motor decodes and makes concrete the actions of the brain. Here you add units, dimensions and goals to the abstract Hal stream from the brain.

There are intermediate nodes between nodes and motors. In the computer analogy, these are drivers. The first is MoNode which converts number into pulsed voltage via an H-bridge. I -> velocity.

There is also an SMNode that converts number into a servo timed pulse.

Motors are in an engineering area where you make whatever is required.

Hal Algebra and Brains

Hal algebra is a system of all this.
Hal objects = {data streams}
Hal actions = {single input nodes, two input nodes}
single input nodes = {d, n,i, not,...}
two input nodes = (+,-, o, a, i, z, g, l, is, gt, lt,...}

HalNodes are one and two input, one output functions.
Out = d(I) is a one input node.
Out = L - R is a two input node. Two input nodes can be connected into n input HalTree functions. HalTrees are many
to one functions.
Hal Algebra can describe and do neural functions.

Book 5
Hal Trees

Hal trees are data stream variables. They are created and operated on by Hal Nodes. They can also be mathematical objects. For example:

As a matrix object

A matrix is a rectangular set of ordered mathematical objects. They are r by c, rows by columns, of elements or objects. Here is a 2 by 2 matrix:

$$a_{11}\ a_{12}$$
$$a_{21}\ a_{22}$$

It takes two subscripts to identify every element.
Now make a triangular matrix like this:

$$a_{11}\ a_{12}$$
$$a_2$$

Replace the matrix objects with MC machines. Now instead of computing on the elements, the elements compute on themselves. Now $a_2 = a_{11}\ O\ a_{12}$ where O is any computer program. Data flows from top to bottom here.

Call the triangular MC matrix T. Then T = F(elements) where F is any computer program in the MCs. Also, call the

triangular MC matrix a HalTree, since I
invented it.

Each HalTree is an element in a higher
level HalTree. That means arithmetic can be
done with HalTrees. HalTrees build HalTrees
to any level. HalTrees map to numbers
exactly.

Let MC elements be nodes.
HalTree = {nodes}
HalTree1 = {HalTrees}
HalTree2 = {HalTree1s}

Let Eij be MC elements.
Then HT = {ij from 00 to mn|Eij}

E11 E12 E13
    E22 E23
        E33
Let that be Tij
Then HT1 = {ij from 00 to mn|Tij}

```
T11 T12 T13
    T22 T23
        T33
```

The properties of trees that are important are: Input size and number of nodes. Let input size be S and number of nodes be N. Let the top row of the matrix be inputs or sensor nodes. Then S is the number of inputs or sensors. There is a single output for every HalTree. The number of nodes is $S + S-1 + S-2 ... + S-(S-1)$.

S	N
1	1
2	1
3	6
4	10
5	15
6	21

. . .

N is the value of the triangular number related to S.

Full and Sparse Trees

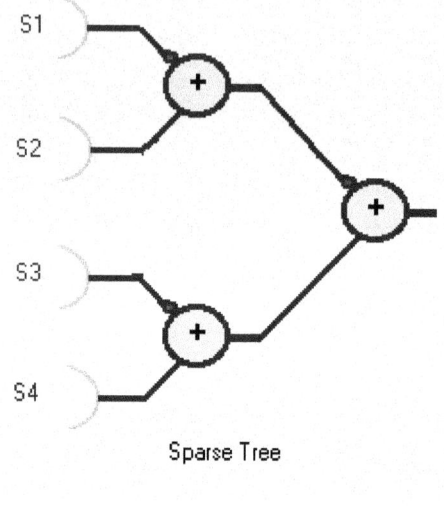

Sparse Tree

O = S1 + S2 + S3 + S4

Figure 34 A sparse PNode tree

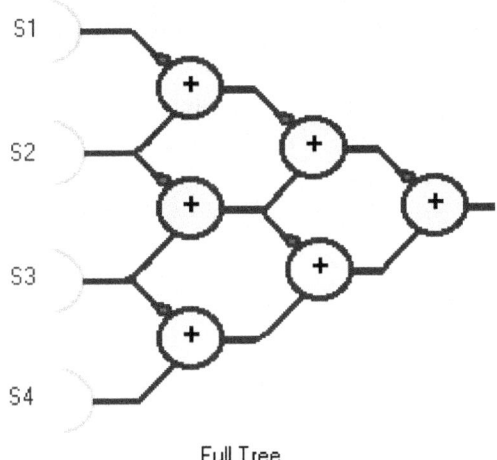

Full Tree

$$O = S1 + 3*S2 + 3*S3 + S4$$

Figure 35   A Full PNode Tree

Full Hal Trees

The output of a full HalTree is an equation
with S terms. The S terms are multiplied by
Pascal tree numbers.

S	Terms
0	a
1	a1 O 1*a2
2	a1 O 2*a2 O 1*a3
3	a1 O 3*a2 O 3*a3 O 1*a4
4	a1 O 4*a2 O 6*a3 O 4*a4 O 1*a5

. . .

Where again, O is any MC function.

Sparse HalTrees

There are also sparse HalTrees where N = S - 1 and all terms have coefficients of 1.

```
a11 a12 a13 a14
a21 a22
a31
```

S = 4 and N = 3 ; 4 - 1.
Each row of a sparse HalTree is 1/2 the number of nodes in the row above it.

S	Terms
1	a1
2	a1 O a2
3	a1 O a2 O a3
. . .	

HalTrees as numbers.

Full HalTrees are N. Each dot is a HalNode or a countable object. * denotes a data input, showing flow direction. o is tree output.

```
               12345
*              * * * * *
* .                . . . .
* . .               . . .
* . . .              . .
* . . . o            o
54321
```

Sparse HalTrees are doubling numbers of HalNodes.

```
                 8421
o           1    *
. .         2    *
. . . .     4    *
* * * * * * * *  8    *
                 *
                 * .
                 * .
                 * . .
                 * . . o
```

The Hal Node functions o and a give the
same result in either full or sparse trees.
P and C nodes are quite different.

Conditional nodes are also different.
The difference is in the type of
computation. Conditional nodes do not make
HalTrees but they work on and in them.

o and a do not compute. They select,
but do not change the data values. If A is
the largest of n inputs, A will be the
output.

Combinatorial Trees

A combinatorial tree is a network of
nodes, many sensors to one motor. The
simplest tree is a single node. F = S1 -
S2, two sensors to one function. When you
add more sensors for more complex trees,
you must also add some intermediate
variables. Data streams sweep though the
tree one layer at a click of c time. Trees
have short-term memory.

The number of sensors can characterize
trees. I find that there are two types of
trees, full trees and sparse trees. A
picture is best to show this. (See figures
38 and 39.) But I need a computational
method also. Let sensors be S1 through Sn.

Then the simplest tree is T2 = S1 - S2. The
next tree is:

```
m1 = S1 - S2 ; note 1 temporary variable,
m1
T3 = m1 - S3   ; 3 input function
```

For 4 inputs:

```
m1 = S1 - S2 ; first temporary variable
m2 = S2 - S3
m3 = S3 - S4

m11 = m1 - m2 ; First second layer variable
m12 = m2 - m3 ;
T4 = m11 - m12 ;
```

Note that a four-sensor tree contains all the trees below it. Those above are full trees. Two input trees cannot tell the difference. They are both full and sparse just as the number two cannot tell addition from multiplication.

The mxx variables are true memory. They persist until the sensor change reaches them.

If I write T as a single equation, eliminating the memory variables, I discover something interesting.

$T3 = m11-m12 = (m1-m2)-(m2-m3) = m1 - 2m2 + m3$; $1 - 2 + 1$
$T3 = m1-2*m2-m3 = S1-S2 - 2*(S2-S3) - (S3-S4)$
Now I have T in terms of S.

$T4 = S1 - 3*S2 + 3*S3 - S4$; Note the coefficients. You see the Pascal triangle under there. This is why I call them combinatorial trees. You can just write down the T function for 5 sensors.

$T5 = S1 - 4*S2 + 6*S3 - 4*S4 + S5$

Given a table of binomial coefficients, you can write down the output weighting of any number of inputs.

You can count the nodes and the memories.

```
Let s = number of sensors or inputs, 2 3 4
5 …
Let N = the counting numbers, … 0 1 2 3 …
Let n = the number of nodes

n = i(N)[s-1] - 0 ; the number of nodes is
the (s-1)th triangular number.

N = … 1   2   3   4   5 …
s =     2   3   4   5 …
n =     1   3   6  10  15 …   ; These are the
triangular numbers.

For 2 sensors, n =  1
For 3 sensors, n =  3
For 4 sensors, n =  6
For 5 sensors, n = 10
and so on.
```

The number of memories is n - 1 because the final output, Ts, is not called a memory.

A tree of nodes retains the character of the single node. In the case of CNodes, the tree has a positive side and a negative side. Feedback to the negative side oscillates. Feedback to the positive side counts, as in the single node. Then the tree has its own combinatorial characteristics.

Tree functions:

If you are not interested in the intermediate memory variables in the tree, you can express trees as functions like this:

O is the output of the tree. The inputs are the sensors S1, S2, S3, S4, … You can have a tree of any node type. (Mixed node type trees I will consider later.) I use a symbol to indicate node type.

Symbol Node
C   CNode
P   PNode
o   ONode
a   ANode

These are the only nodes I will consider here. All tree functions are the same for all these, so I will use only C type below. Understand that any of the above types can be substituted.

Then I can express the function:

```
OC2 = CTree(S1,S2) = S1 - S2
OC3 = CTree(S1,S2,S3) = S1 - 2*S2 + S3
OC4 = CTree(S1,S2,S3,S4) = S1 - 3*S2 + 3*S3 - S4
OC5 = CTree(S1,S2,S3,S4,S5) = S1 - 4*S2 + 6*S3 -
                                        4*S5 + S5
```

and so on.

If you set all the sensors to 1 and use PNodes you get the series:

```
OP1 = PTree(1) = 1
OP2 = PTree(1,1) = 1 + 1 = 2
OP3 = PTree(1,1,1) = 1 + 2 + 1 = 4
OP4 = PTree(1,1,1,1) = 1 + 3 + 3 + 1 = 8
OP5 = PTree(1,1,1,1,1) = 1 + 4 + 6 + 4 + 1
= 16
OP6 = PTree(1,1,1,1,1,1) = 1 + 5 + 10 + 10
+ 5 + 1 = 32
```

Let SN = number of sensors.
Let SC = sum of coefficients

$$SC = 2^{(SN-1)}$$

I will rewrite the coefficients so they line up better.

C1	C2	C3	C4	C5	C6	C7	C8	C9		
1									=	1
1	+	1							=	2
1	+	2	+	1					=	4
1	+	3	+	3	+	1			=	8
1	+	4	+	6	+	4	+	1	=	16

...

And so on.

C2 is N and is the number of wires required to carry the sum on the right. C3 contains the triangular numbers and the rest of the columns contain combinatorial numbers.

Each column in this display is the integral of the column before it.

C2 is i(C1) and C3 is i(C2) or in general:
C(n) = i(C(n-1)) - 0

So we have established the primary, underlying arithmetic of trees. Now let's go back to CNode trees and try some other inputs.
Input the beginning of the triangular numbers, just for kicks. Here are the coefficients.

```
2: 1 - 1
3: 1 - 2 +  1
4: 1 - 3 +  3 -  1
5: 1 - 4 +  6 -  4 +  1
6: 1 - 5 + 10 - 10 +  5 -  1
7: 1 - 6 + 15 - 20 + 15 -  6 +  1
8: 1 - 7 + 21 - 35 + 35 - 21 +  7 - 1
9: 1 - 8 + 28 - 56 + 70 - 56 + 28 - 8 + 1
```

OCn = CTree(S1,S2,S3,S4,S5,...n)
OC2 = CTree(1,3) = 1*1-3*1 = -2
OC3 = CTree(1,3,6) = 1*1 - 3*2 + 6*1 = 1
OC4 = CTree(1,3,6,10) = 1*1 - 3*3 + 6*3 - 10*1 = 0
OC5 = CTree(1,3,6,10,15) =1***1** -3***4** + 6***6** - 10***4** + 15***1** = 0

```
OC4 = CTree(1,2,2,1) = 1*1-2*3+2*3-1*1 =    0
OC4 = CTree(1,1,1,1) = 1*1-1*3+1*3-1*1 =    0
OC4 = CTree(1,3,1,3) = 1*1-3*3+1*3-3*1 =   -8
OC4 = CTree(1,4,1,4) = 1*1-4*3+1*3-4*1 =  -12
OC4 = CTree(4,1,4,1) = 4*1-1*3+4*3-1*1 =   12
OC4 = CTree(3,1,4,1) = 3*1-1*3+4*3-1*1 =    6
```

This is what a full tree looks like in a spreadsheet:

```
      Input mem1 mem2 Output
      A     B    C    D
Row1  S1
Row2  S2    M11
Row3  S3    M12  M21
Row4  S4    M13  M22  OC4
```

The equations are:
M11 = S2 - S1   ;This column differentiates the input
M12 = S3 - S2
M13 = S4 - S3

M21 = M12 - M11 ;M2x = d(M1x)
M22 = M13 - M12
OC4 = M22 - M21 ;OC4 = d(M2x)

You can write every full tree in this
format, which shows some of its
relationships. The numbers are: Column and
row. This tree is made of CNodes. This is
also a spreadsheet format.

```
s11
s12   m21
s13   m22   m31
s14   m23   m32   m41
s15   m24   m33   m42   m51
```

With CNodes, m2x = d(s1x) and m3x = d(m2x)
and so on.

Let's put in some numbers. First N.

```
1
2    1
3    1    0
4    1    0    0
5    1    0    0    0
```

Now doubling numbers:

```
1
2    1
4    2    1
8    4    2    1
16   8    4    2    1
```

Now tripling numbers:

```
1
3    2
9    6    4
27   18   12   8
81   54   36   24   16
```

Note that all the diagonal numbers are doubling numbers.

Let's write down another tree with the differences in the rows from column to column.

```
1                        ;3 - 2
3    2                   ;9 - 6 and 6 - 4
9    6    4              ; and so on
27   18   12   8
81   54   36   24   16
```

How is that for being fractal?

How about putting some combinatorial numbers in a CTree like this:

```
1
6    5
21   15   10
56   35   20   10
126  70   35   15   5
252  126  56   21   6    1
```

Put the diagonal in a PTree.

```
1
5    6
10   15   21
10   20   35   56
5    15   35   70   126
1    6    21   56   126  252
```

And there is a relation between combinatorial numbers and binomial numbers, all done mechanically.

Sparse trees

Again, use sensors S1 to Sn.

A two input tree:

T = S1 - S2

Three input sparse tree:

m1 = S1 - S2
T = m1 - S3

Four inputs:

m1 = S1 - S2
m2 = S3 - S4
T = m1 - m2

    Again, eliminating the memory variables, you obtain:
T = S1 - S2 + S3 - S4 ; All the coefficients are 1.
    Sparse trees sample the inputs, meaning some inputs will be missed. They are, in that sense, an approximation of the inputs and their outputs will usually differ from the outputs of a comparable full tree.
    Since all the coefficients of sparse trees are 1, you can use them where no center bias or weighting is required.

The function CSTree(S1,S2,S3,S4)
is S1 - S2 + S3 - S4

(other node types)	P	o	a
OC2 = CSTree(1,2) = -1	3	2	1
OC2 = CSTree(2,3) = -1	5	3	2
OC3 = CSTree(1,2,3) = 2	6	3	1
OC3 = CSTree(2,3,4) = 3	9	4	2
OC3 = CSTree(3,4,5) = 4	12	5	3
OC4 = CSTree(1,2,3,4) = -2	10	4	1
OC4 = CSTree(2,3,4,5) = -2	14	4	2
OC4 = CSTree(3,4,5,6) = -2	18	6	3
OC4 = CSTree(0,7,10,7) = -4	24	10	0
OC4 = CSTree(7,10,7,0) = 4	24	10	0
OC5 = CSTree(10,7,0,7) = -4	24	10	0
OC5 = CSTree(1,2,3,4,5) = 3	15	5	1
OC5 = CSTree(2,3,4,5,6) = 4	20	6	2
OC5 = CSTree(3,4,5,6,7) = 5	25	7	3

———

Trees with mixed node types

As you have seen, trees of the same type compute combinatorial numbers. OT = CTree(a,b,c,d) If you cycle through the values of {a,b,c,d} you get zeroes for all equal or linear values but you lose direction. The tree is symmetric and weighted toward the center. The last node that feeds the motor always gives direction and if you make that a CNode, you retain direction if the two sides of the node are collected for it. For example:

```
let OT = Left - right;
Then if Left = a + b
and     Right = c + d
```

Then you know which side is bigger, so you know direction.

From arithmetic, you know that the sum of the differences equals the difference of the sums. So you get exactly the same result if OT = Left + Right
and Left = a - b
and Right = c - d

This also works:
```
OT = Left - Right
Left = a o b
Right = c o d
```

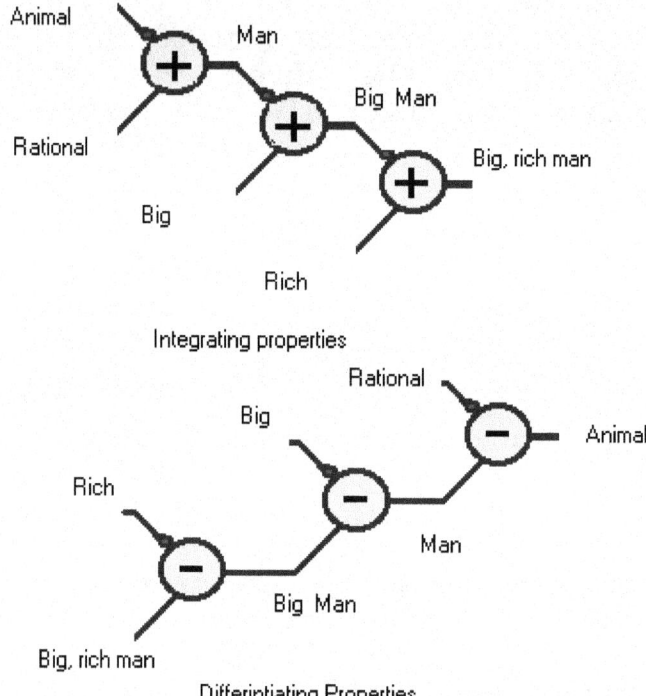

Integrating properties

Differintiating Properties

Figure 36   Property Trees

226

Property Trees

A property tree is the outer diagonal of a regular full tree. Property trees are of the form:

O2 = O1 + a1
O3 = O2 + a2
O4 = O3 + a3
And so on.

And:
O3 = O4 - a3
O2 = O3 - a2
O1 = O2 - a1

The reason I call these property trees is shown by a change in variable name.

Integration of a concept:
man = rational + animal
big man = man + big
rich, big man = big man + rich
and so on.

Also, differentiation of a concept:
big man = rich, big man - rich
man = big man - big
animal = man - rational
And so on.

Feedback

While there is some memory in tree networks, permanent memory is in feedback loops and special structures. Here you leave the computer and go into uncharted territory. Tree networks can be run on spreadsheets, but feedback loops break computers.

Direct feedback of the type O = O - R and O = I - O can be done with nodes easily. You can characterize the action of most nodes. In general:

Node	I	R	O	action
CNode	O	R	O	O = O - R Counts up or down
CNode	I	O	O	O = I - O Oscillates I to O
PNode	I	O	O	O = I + O Counts up or down
PNode	O	R	O	O = O + R Counts up or down
ONode	I	O	O	O = I o O Locks in highest I
ONode	O	R	O	O = O o R Locks in highest R
ANode	O	R	O	O = O a R Locks in lowest R
ANode	I	O	O	O = I a O Locks in lowest I

Natural feedback from actions is not direct. Generally, there will be some divisor of the action and a time delay. For the general node:

how = what - where ; you usually provide a sensor on **where** that relates the action from **how**. This provides negative feedback so the process tends to center, with **how** = 0.

If you reverse **what** and **where** so you have:

how = where - what ; then you will tend to lock up at some maximum or minimum **how**.

Feedback is extremely context sensitive. It matters greatly how the motor and sensors are arranged. If you are controlling some action, you must arrange the motor and sensors so that a zero from the node indicates the stopping point. For example, say you are controlling the speed of an engine. You measure the speed and send it to the R input of a CNode. You have a dial, marked as speed, (RPM) that produces a data stream that you connect to the I input. You want to obtain this RPM. The O output of the CNode is connected to the motor that in turn is connected to the throttle. When the motor is still, the throttle does not move. When the motor turns, the throttle moves so that RPM is increased with positive rotation and decreased with negative rotation.

Now, after all this, you have a little servo. You can dial in an RPM and the engine will follow. The reason for the extra motor for the throttle is that the output of the CNode is the change in RPM needed. The change is used to set the throttle position, so the throttle integrates the output of the motor. The position of the throttle produces a fixed RPM from the engine, changing position to speed. That RPM is compared to the RPM you have dialed in, and the position of the throttle is moved until the two RPM measurements are equal, no matter what.

With feedback, you have to pay attention to units. You may have to integrate or differentiate a data stream to make the units match. Sometimes, as in our example, the hardware itself performs the unit change. Nature is full of integrators and differentiators. Fortunately, most of them are simple changes of position to velocity and back.

Numbers as Data Streams

Data streams can be any mathematical function, including streams of number sequences.

Let **N** = … **0 1 2 3 4 5 6 7 8 9 10** …

N = i(1)   where 1 could be 1 bit.

d(N) = 1

These are the counting numbers. You know they will be a type 2 data stream if read successively. The ancients would have said this data stream could go on forever, but you know better, don't you. Every data stream is a wheel, with these counting numbers as the click. So under every other number are these counting numbers. N, as a data stream, relates exactly to 1 bit in its arithmetic. As 1 bit cannot distinguish between plus and minus operations, N cannot distinguish offset and phase shift.

Note that the scale of **N** does affect its derivative. That is: **kN =   … 0k 1k 2k 3k 4k** … has the derivative **k**.
    d(kN) = k = k*d(N)
kN = i(k)
d(kN) = k

Of course, if I could divide, then **kN/k = N**.

Sometimes we, as mathematicians, need to refer to particular numbers in the data stream. Generally, we have and need access to any two successive numbers. We can use **i** as an index into the data stream so we can make statements like:

d(kN)[i] = kN[i] - kN[i-1] which we recognize as differentiation, from the definition.

i, as an index, is a counting number stream with **d(i) = 1**.
Any and every data stream has an implicit **i** stream associated with it.

All **kN** streams are linear ramps. **k** puts space between the numbers in this way:

k = 1, N = … 0 1 2 3 4 5… Showing every number. That is, **N** is the same as **i**, the index. **N[i] = i**

When **k** is greater than 1, then **i** does not match **N**.

```
N  = … 0 1 2 3 4 5 6 7 …
i  =   0 1 2 3 4 5 6 7 …

2N = … 0 2 4 6 8 10 12 14…
```

```
i  =    0 1 2 3 4 5  6  7 …
```

Since **kN/k = N** every type 2 data stream can
be reduced to a counting number stream.

Power Streams

Mathematicians have been enamored of power series since they learned to multiply. Take N and make powers of N and you have a power data stream. Let Nn represent N^n for convenience.

```
N0 = 1   1   1   1    1...
N1 = 0   1   2   3    4    5    6...
N2 = 0   1   4   9   16   25   36...
N3 = 0   1   8  27   64  125  216...
N4 = 0   1  16  81  256  625  ...
```
and so on.

Here N2 does not equal d(N3) except for N1 and N0. We can differentiate any Nx and by doing so you will find a generator for any Nx and we eliminate x.

Going Down

```
N4  =   0   1   16   81  256  625...   = x⁴
d(N4) =     1   15   65  175  369...   = 4x³
d2(N4) =        14   50  110  194...   = 12x²
d3(N4) =             36   60   84...   = 24x
d4(N4) =                  24   24...   = 24
```

$$N4 = x^4$$
$$d(N4) = 4x^3$$
$$d2(N4) = 12x^2$$
$$d3(N4) = 24x$$
$$d4(N4) = 24$$

```
N3  =    0 1   8   27   64  ...
d(N3) =    1   7   19   37   61   91...
d2(N3) =       6   12   18   24   30...
d3(N3) =           6    6    6    6...
```

```
N2  =     0   1   4   9  16  25
d(N2)  =      1   3   5   7   9 ...
d2(N2)  =         2   2   2   2

N  =   0   1   2   3   4   5   6...
d(N)  =   1   1   1   1   1   1...
```

Observe that $dn(n\wedge n) = n!$

Going Up

Generating powers of N streams is done by these rules:

N = i(1) start at 0

N2 = i2(2) start at 1

N3 = i3(6) start at 0

N4 = i4(24) start at 12

In general:
Nx = ix(x!) starting at x!/2 if x even else starting at 0

For example:
N1 = i1(1) starting at 0
N2 = i2(1*2) starting at (1*2)/2
N3 = i3(1*2*3) starting at 0
N4 = i4(1*2*3*4) starting at 4!/2 ; 4! = 1*2*3*4
N5 = i5(5!) starting at 0
N6 = i6(6!) starting at 6!/2

So:
Nx = ix(x!) starting at x!/2 if x even,
else start at 0

N1=i1(1) = 0   1   2   3   4   5   6   7   8
9   10   11 …
N2=i2(2) =   1   4   9   16   25   36   49   64
81   100   121 …

```
N3=i3(6) = 0  1  8  27   64 125 216 343 512
729 1000 1331 …
N4=i4(24) =   1 16  81  256 625 …
N5=i5(120)=0  1 32 243 1024 …
N6=i6(720)=   1 64 729 4096 …
```

Showing integration with INodes and CNodes, you have:

```
N1 = i(1) - 1 =           0  1   2   3  4    5   6 …
N2 = i(i(2) - 1) - 0 =    0  1   4   9 16   25  36 …
N3 = i(i(i(6)-6)-0)-0 = 0  1   8  27 64 125 216 …
```

with power notation.

$$N1 = i(1) - 1 = \qquad\qquad 0^1 \quad 1^1 \quad 2^1 \quad 3^1 \quad 4^1 \quad 5^1 \quad 6^1 \ldots$$
$$N2 = i(i(2) - 1) - 0 = \quad 0^2 \quad 1^2 \quad 2^2 \quad 3^2 \quad 4^2 \quad 5^2 \quad 6^2 \ldots$$
$$N3 = i(i(i(6)-6)-0)-0 = 0^3 \quad 1^3 \quad 2^3 \quad 3^3 \quad 4^3 \quad 5^3 \quad 6^3 \ldots$$

Power data streams are a special case of
polynomial data streams. Remember:

$$1111 = 1*b3 + 1*b2 + 1*b + 1$$

If b = 0 then 1111 = 1.
If b = 1 then 1111 = 4.
etc.
As a table:

b	D3	D2	D1	D0	=	R	dR	ddR	dddR
0	0 +	0 +	0 +	1	=	1			
1	1 +	1 +	1 +	1	=	4		3	
2	8 +	4 +	2 +	1	=	15	11	8	
3	27 +	9 +	3 +	1	=	40	25	14	6
4	64 +	16 +	4 +	1	=	85	45	20	6
5	125 +	25 +	5 +	1	=	156	71	26	6
6	216 +	36 +	6 +	1	=	259	103	32	6
7	343 +	49 +	7 +	1	=	400	141	38	6
8	512 +	64 +	8 +	1	=	585	185	44	6
9	729 +	81 +	9 +	1	=	820	235	50	6
10	1000 +	100 +	10 +	1	=	1111	291	56	6

As data streams:
D0 = 1 1 1 1 1 1 1 1 1 1 …
D1 = 0 1 2 3 4 5 6 7 8 9 …

D2 = 0 1 4 9 16 25 36 49 64 81 100…
d(D2) = 1 3 5 7 9 11 13 15 17
d(d(D2)) = 2 2 2 2 2 2

D3 = 0 1 8 27 64 125 …
d(D3) = 0 1 7 19 37 61
d(d(D3)) = 6 12 18 24 30…
d(d(d(D3))) = 6 6 6 6 6 6 6 6…

For the general polynomial a2*b2 + a1*b1 + a0
multiply D0 by a0, D1 by a1 and D2 by a2.

The instructions for making the first digit of any polynomial number is to take a constant, a0, and integrate it one time.

For the second digit take a1*2 and integrate it twice.

For the third digit take a2*6 and integrate it three times.

For the nth digit take an*n! and integrate it n times.

Here is a table of the instructions.

Digit	Coefficient	Multiplier	Number of integrations
1	a0	1	1
2	a1	2	2
3	a2	6	3
4	a3	24	4
5	a4	120	5
...			

Manipulating Time and Offset in data streams

The data stream passing through any node has a present time, a past time and a future time. It takes a time of c to shift one to the other through the node. There is another way to shift time in the data stream as I have noted previously. Simply adding or subtracting a lower degree data stream can shift some data stream types in time, forward or backward.

Constant data streams cannot be shifted, since they are the same at all times. Adding or subtracting a constant can offset them.

Polynomials can be shifted or offset. The rule for time shifting a polynomial is to add or subtract its derivative. Anything else will simply be an offset.

The polynomial of degree 1 (N) cannot be offset. Adding a constant is adding a derivative and anything added or subtracted is a shift in time. In order to shift a data stream in time you must know its derivative. If you have the derivative data stream, you know how to be anywhere in time on the data stream created from the derivative. By adding or subtracting the derivative to the polynomial data stream, you shift it upward or backward in relative time. Only the derivative will do this. If the data stream has no derivative that is ultimately a constant, then the data stream

is not a polynomial and what you are adding
or subtracting is an offset.

Some examples:

A constant is the same everywhere. Its
derivative is zero.

N = {0 1 2 3 4 5 6 …} has a derivative of
one.

Anything you add or subtract from an N data
stream shifts it in time. N = {0 1 2 3 4 5
6 …}
```
              +   {1 1 1 1 1 1 1 …}
              =   {1 2 3 4 5 6 7 …} ; we
```
shift forward 1 click.

```
         2N = {0 2 4 6  8 10 …}
         +2   {2 2 2 2  2  2 …}
         =    {2 4 6 8 10 12 …} ; We shift
```
up 1 click.

   For any kN we use k to shift 1 click.
Use nkN to shift n clicks.
   The same applies to higher degree
polynomials, but now the derivatives are
not constant. To properly shift a higher
degree polynomial in time you must start
adding derivatives at the proper point so
the derivatives line up with the
polynomial.

Regular mathematical notation is awkward here, so I do this:

P(x)[j] is the jth number of the xth integral of P.

To shift such a data stream in time make use of the relation:

P(x)[j] = P(x-1)[j] + P(x)[j-1] ; In other words a data stream of x degree is built by adding a data stream of x-1 degree to the past of the x degree data stream. That means that every polynomial carries its own derivative, which means that every polynomial can be differentiated at least once.

Also: P(x) = P(x-1) + P(x-2) … + P + k

Comment and Closure

There are several reasons I have for promoting this. I think it has some novelty and newness. The very idea of replacing functions with computers changes mathematics in ways we cannot predict. Now we have self computing structures and we have expanded the size of the computing set itself by adding conditional arithmetic and arithmetic functions with memory.

References

Ashby, W. Ross, Design for a Brain, Wiley & Sons, 1960
Stability. The homeostat. Cybernetics. Early IF -THEN not adequate discovery. This started control systems. No one paid attention to the limitations.

Bell, Eric Temple, (1883-1960) The Magic of Numbers, Dover, 1997 (reprint of McGraw-Hill, 1946)
The problems of Pythagoras in the invention of the scientific method and the magic of numerology at the same time.

Cardwell, Donald, The Norton History of Technology, W.W. Norton & Company, 1995
A good history of technology in context with general history. He points out in several places that new technology is always produced by outsiders. I would extend this to all new ideas.

Cohen, Jack and Stewart, Ian, The Collapse of Chaos, Viking, 1994
Simplicity and complexity. Simplexity and complicity. How a complex world can appear simple.

Kelso, J. A. Scott, <u>Dynamic Patterns</u>, The MIT Press, 1995
Self-organization and patterns. No mechanism yet.

Kline, Morris, <u>Mathematics and the Search for Knowledge</u>, Oxford University Press, 1985
What mathematics reveals about nature.

Levenson, Thompson, <u>Measure for Measure</u>, Simon & Schuster, 1994
"A Musical History of Science." This is a beautiful essay on science, from the Pythagorean beginnings.

Lines, Malcom, <u>A Number for Your Thoughts</u>, IOP Publishing, Ltd. 1988
A book about numbers that starts with counting. Like all such books, there is much ado about prime numbers, but, on the whole, it is an entertaining book.

Mandelbrot, Benoit B, <u>The Fractal Geometry of Nature</u>, W.H. Freeman and Company, 1977,1983
This is the book that told us the world was fractal. Before that we did not think about it because we could not see it.

McNiel, Daniel and Freiberger, Paul, <u>Fuzzy Logic</u>, Simon & Schuster, 1993
Story of fuzzy development.

Milne, W.E., Numerical Calculus, Princeton
University Press, 1949
Probably out of print. This reference was in
Van Nostrands Scientific Encyclopedia under
"difference". There is a whole field of
discrete or digital mathematics that
parallels analog mathematics.

Reed, Harold L., Brains for
Machines/Machines for Brains, Nova Science
Publishers, Inc., 1996
Hal network nodes and algebra are developed
here in a story of discovery. We look for
the least brain.

Schroeder, Manfred, Fractals, Chaos, Power
Laws, W.H. Freeman and Company, 1991
Good stuff on self similarity and scale
invariance, which I use eagerly.

Index

BIO

I am Harold L. Reed, the author of
<u>Brains for Machines/ Machines for Brains</u>,
Nova Science, 1996 and the inventor and
developer of the mathematics and hardware
of this system. <u>Brains for Machines</u> was
about discovering the least brain.
A new book is underway to extend that book.
Working title is <u>Neurons for Robots/ A Tool
Set</u>.
I am happy to discuss this system with any
serious person. Everyone is an amateur in
this system and it is upside down and
backwards from conventional thought enough
so that some learning is required.
hlreed@halbrain.com
http//www.halbrain.com

www.ingramcontent.com/pod-product-compliance
Lightning Source LLC
Chambersburg PA
CBHW071409170526
45165CB00001B/227